Trust in Co-operative Contracting in Construction

Trust in Co-operative Contracting in Construction

Sai On CHEUNG

City University of Hong Kong Press

First published 2007

ISBN: 978-962-937-117-3

Photos: Courtesy of Raymond Wai Man Wong

Published by:
    City University of Hong Kong Press
    Tat Chee Avenue, Kowloon, Hong Kong
    Website: www.cityu.edu.hk/upress
    E-mail: upress@cityu.edu.hk

Printed in Hong Kong

Contents at a Glance

Detailed Chapter Contents

Foreword

BY PROFESSOR CLIFF HARDCASTLE

The importance of the construction industry to the health and welfare of society cannot be overstated. It is therefore somewhat surprising that it is only in the last half century that its performance has received the close attention of policy makers, practitioners and researchers seeking to ensure that the significant resources invested in it achieve the desired outcomes in an efficient and effective manner. The early work of these stakeholders concentrated on the efficient use of labour, technical and process innovation and alternative procurement approaches. Relationships, both inter-personal and inter-organisational were rarely commented upon and fundamental recommendations did not follow from an understanding of the importance of relationships in the delivery of construction projects.

More recently, as a consequence of an increasing realisation among all stakeholders that it is the people and the dynamics among them which are key to the successful delivery of construction projects, policy makers have begun to make recommendations to improve the quality of relationships and practitioners have sought to evolve contractual and communication systems which would lead to improved relationships. On the other hand, researchers have sought to understand and describe the issues. In the near past this has lead to a surfeit of publications seeking to contribute to the creation of the "perfect" contractual model and the "paradigm shift" which may lead to it. Fundamental to the establishment of high quality, high value relationships is the principle of trust, a concept often ill defined and contextually specific.

Dr. Cheung has devoted much of his career to the exploration of trust in relationships within the context of the construction industry. The result is a wealth of knowledge and expertise founded upon rigorous application of research methodology. It is precisely these attributes that make this book an invaluable resource for those seeking to move the industry forward. In order for a paradigm shift to take place within the industry it is not enough for policy makers to propound its merits or even to simply recognise the benefits of trust in contracting relations, but also for stakeholders to feel comfortable that these benefits will be manifest when trust is established. This requires evidence and well founded argument. Dr. Cheung's work contains both and will make a valuable contribution to changing the manner, tone and substance of relationships in the industry in the future.

Professor Cliff Hardcastle
Deputy Vice Chancellor
University of Teessside
United Kingdom
May 2007

Foreword

BY PROFESSOR ROGER FLANAGAN

Construction around the world is big business; in 2006 the annual output was estimated at US$4.6 trillion. It is one of the most important industries and touches everyone's life in some way by providing homes, infrastructure, schools, hospitals, workplaces and the built environment necessary for survival. It creates wealth, provides jobs, and sustains economic growth.

Despite its importance and size, the industry often comes under criticism and scrutiny, with many countries conducting industry reviews at regular intervals. Interest is focused on improving delivery and the method of procurement, design, production and the types of contract that deliver projects on time and budget. There is a common thread that runs through most of the industry reports: the lack of trust between the parties; the need to work collaboratively; the desire for a fair profit reflecting the risk and uncertainty; and removing conflict.

Clients need projects on time and on budget, contractors need to make a profit and a sensible return on their capital and equity, but in reality it is people that deliver projects and make things happen. People's expectations and beliefs have to be matched by their behaviour. If people and organisations share the same goals, are honest about risk and uncertainty, and create an environment of trust, then much of the conflict can be removed.

This book is a welcome addition to the debate on how the industry can improve trust and collaborative working. The paradigm shift in contracting culture is highlighted with a number of issues being discussed including the spirit of partnering, mutual trust, integrity, openness, teamwork and problem resolution.

The beauty of this book is that it is well researched, well referenced and well written. It takes a broad perspective that is applicable around the world. It challenges the industry to think about its spirit of trust and relationships and aims to put the handshake back into construction.

There is no shortage of innovative methods of procurement and new contracts; there is a shortage of trust.

Professor Roger Flanagan
President of the Chartered Institute of Building
May 2007

Preface

The construction industry is infamous for its adversarial culture. Several key industry-wide reviews all pointed to the need to have a revolutionary reform in the way construction facilities are procured. The situation had become more acute as the construction sector, once regarded as a local industry, is now facing unprecedented global challenges. The procurement paradigm has changed significantly, particularly in the last decade. Collaborative types of endeavor have become the norm. More and more projects are employing integrative arrangements that include finance, design, construction and operation. This changing face of procurement calls for a new mode of contracting behavior—co-operative contracting, which in its broadest sense, is requiring the contracting parties to adopt a co-operative attitude thereby maximizing the synergistic effect. Contracting behavior is typically legally regulated. Moreover, founding a legal anchor for the obligation to co-operate in common law system is discouraging. In fact, the legal profession has been swift to express reservations on the binding effect of the pledge for honestly and co-operation. Nonetheless, the relational contracting theory advocated by a group of U.S. legal scholars has provided a nice support to co-operative contracting within a civil law framework. Nevertheless, this conceptualization remains controversial ever since its first introduction in the mid-seventies. As such, relational contracting theory remains an academic pursuit as the common law system has yet to establish a general doctrine of good faith. Seemingly looking into the behavior aspects in contracting would be more fruitful. In construction, partnering is the closest delivery approach resembling co-operative contracting. Success stories have been reported, one notable pioneer project in Hong Kong is the Mass Transit Railway Tseung Kwan O extension. A case study on one of the contracts of the TKE project is used to demonstrate the importance of behavioral change in co-operative contracting, the partnering arrangement, tools as well as the lessons derived. Trust has been identified as the most critical driver for co-operative behavior. Two key questions need to be addressed. Firstly, among the various forms of trust, which form will most amenable to construction contracting. Secondly, between the contractor and the client, who is the best person to kick start the trust cycle. Trust has been a topical research area as it underpins many of our behavior. However, in engineering and construction fields, where technical skill and knowledge dominate, trust has not been accorded the level of attention that it deserves.

The construction industry is undergoing changes in moving towards a co-operative paradigm in contracting. Chapter one outlines the factors leading to the paradigm shift towards co-operative contracting. Chapter two explores the theoretical bases for co-operative contracting. Operationalizing these theoretical constructs in construction contracts is first given in Chapter three. The importance of trust in co-operative contracting is highlighted in a case study on MTRC TKE project presented in Chapter four. The effects of trust on contracting behavior are analyzed further in Chapter five. Critical trust factors in the views of contracting parties are evaluated in Chapter six. Chapter seven details a study that aimed to identify the party best in initiating a trust cycle in contracting. Lastly, the notion of good faith in the context of co-operative contracting is critically reviewed in Chapter eight. This book provides a comprehensive discussion on trust in co-operative contracting. Directed from both management and legal perspectives with findings supported by empirical researches, it is hoped that professionals, researchers and students will find the book interesting.

Sai On Cheung
City University of Hong Kong
May 2007

List of Illustrations

Tables

Figures

List of Contributors

SAI ON CHEUNG   LLB(Hons), LLM, MBA, MSc (IT Mgt), PhD, FRICS, FHKIS, AAIQS, RPS(QS), MHKIE, MCIOB, MCIArb, RPE(Bldg)

Dr Cheung is experienced in construction contract and dispute management. Building on his experience in both consultancy office and contracting organizations before joining the City University of Hong Kong in 1989, he has developed a research portfolio that addresses some key issues in construction dispute management. These include dispute avoidance, design of dispute resolution processes, negotiation and mediation of construction disputes. In 2002, he established the Construction Dispute Resolution Research Unit. He received two awards from the Chartered Institute of Building for his collaborative research work with the industry on partnering and control of variations. Dr. Cheung has more than one hundred and fifty publications mainly in international refereed academic journals. He also serves on the editorial board of six international refereed academic journals and commission.

HENRY C. H. SUEN   BArch (Hons), MSc, LLB (Hons), Barrister (Gray's Inn).

Mr. Suen studied architecture in Australia and construction project management in Hong Kong. He read law at the London University and was called to the English Bar in 2005. He is a Barrister of Gray's Inn. Before joining the Department of Building and Construction, he had worked in the construction industry for five years with hands-on experience in construction project management and buildings submissions. His teaching, research interest and publication are in the areas of construction law.

PETER SHEK PUI WONG   BSc, MPhil, PhD, AIQS (Affil.), MIEEE

Dr. Wong obtained his MPhil and PhD in supply chain management from the Department of Building and Construction, City University of Hong Kong. He is an affiliate member of the Australian Institute of Quantity Surveyors (AIQS) and a member in Engineering Management division of the Institute of Electrical and Electronics Engineers (IEEE). Before embarking on his research career, he had worked as a quantity surveyor in a major cost consultant office and involved in a number of prestigious construction projects, including the Development of the II International Finance Centre in Hong Kong. Dr. Wong has considerable research experience in construction management related topics, such as: organizational learning, value chain and construction partnering. He is a member of the Construction Dispute Resolution Research Unit.

WEI KEI WONG   BSC, MPHIL

Miss Wong obtained her BSc in Quantity Surveying and MPhil in Construction Management from the City University of Hong Kong. She is currently a research assistant of the Construction Dispute Resolution Research Unit. Her principal research is in developing a trust inventory for use in the construction industry in Hong Kong.

TAK WING YIU   BSC, PHD, AIQS (AFFIL.)

Dr. Yiu teaches in the Department of Building and Construction, City University of Hong Kong. He is an associate director of the Construction Dispute Resolution Research Unit. His current research interests are in dynamic modeling of construction dispute negotiation and mediation. Notable examples include investigating the dynamic change in dispute resolution behavior and contingent use of mediator and negotiator tactics. His doctoral thesis, entitled "A Behavioral Analysis of Construction Dispute Negotiation", won the 2006 Hong Kong Institute of Surveyors (HKIS) dissertation (PhD category) award.

Trust in Co-operative Contracting in Construction

1

Paradigm Shift in Contracting Culture

Sai On Cheung

1.1 Construction Industry and the Hong Kong Economy

The construction industry is a vital part of the economy of many major cities in the world. Because of its unique position in the world's economy, the industry is simply too important to be allowed to stagnate.

Table 1.1   Contribution of the Construction Industry to GDP in Hong Kong

Year	GDP (HK$million) in Construction Industry in Hong Kong	GDP (HK$million) Overall in Hong Kong	Percentage of Contribution
1980	8,846	143,402	6.5%
1981	12,259	172,965	7.3%
1982	13,205	195,408	7.1%
1983	12,729	216,383	6.2%
1984	12,782	260,761	5.2%
1985	12,551	276,823	4.8%
1986	14,118	319,232	4.7%
1987	16,853	393,541	4.5%
1988	20,140	465,245	4.5%
1989	25,331	536,268	5.0%
1990	29,701	598,950	5.2%
1991	33,915	690,324	5.2%
1992	36,528	805,082	4.8%
1993	42,177	927,996	4.9%
1994	45,356	1,047,470	4.6%
1995	53,694	1,115,739	5.1%
1996	64,115	1,229,481	5.4%
1997	71,190	1,365,024	5.5%
1998	69,101	1,292,764	5.7%
1999	65,560	1,266,702	5.5%
2000	62,054	1,314,789	4.9%
2001	57,167	1,298,813	4.6%
2002	51,534	1,276,757	4.2%
2003	44,910	1,233,983	3.7%
2004[#]	40,376	1,291,568	3.2%
2005[#]	38,612	1,382,052	2.9%

Source: Census and Statistics Department, Hong Kong, 2007
Provisional figure as of January 2007

Table 1.1 shows the contribution in percentage of the construction industry towards the Gross Domestic Product (GDP) of Hong Kong. It can be noted that, despite a drop was experienced in 2003 to 2005, the contribution of the construction industry had been significant and around 5% to 6% of the overall GDP in Hong Kong.

The importance of the construction industry can also be illustrated by the number of persons directly engaged in construction related establishments. Table 1.2 presents such information from 1981 to 2004. Evidently since the early nineties, there have been persistently over two hundred thousand people, accounting to approximately 6–7% of the total workforce, working in the construction related sector.

1.2 Characteristics of the Construction Industry

A number of research studies have investigated the characteristics of the construction industry. Some of the typical characteristics are:

(1) *Diversity of product*—the product of the construction industry includes building of new structures, modifications of existing ones, maintenance, repair, and improvements of houses, apartments, factories, offices, schools, roads, and bridges etc. (Dunican 1985, Bureau of Labor Statistic, U.S. Department of Labor 2005a).

(2) *Segmentation*—the construction industry is divided into three major segments: (i) Construction buildings contractors, or general contractors, build residential, industrial, commercial, and other buildings; (ii) Heavy and civil engineering construction contractors build sewers, roads, highways, bridges, tunnels, and other projects and (iii) Specialty trade contractors performing specialized activities related to construction such as carpentry, painting, plumbing, and electrical work (Bureau of Labor Statistic, U.S. Department of Labor 2005a).

(3) *Long working hour*—most employees in this industry work full time, and many work over 40 hours a week. In the US in 2004, about 1 in 5 construction workers worked 45 hours or more a week. Construction workers may sometimes work evenings, weekends, and holidays to finish a job or take care of an emergency (Bureau of Labor Statistic, U.S. Department of Labor 2005b).

Table 1.2   Labor Force Distribution of Construction Industry in Hong Kong

Year	Number of Persons Directly Engaged (no.) in Construction Related Establishments in				Total Labor Force (no.)	Accounted Percentage to Total Labor Force
	Building and Civil Engineering	Architectural, Surveying and Project Engineering	Real Estate Development, Leasing, Brokerage and Maintenance Management	TOTAL		
1981	114,797	9,737	N.A.	N.A.	N.A.	N.A.
1982	106,711	10,268	N.A.	N.A.	2,498,100	N.A.
1983	96,639	9,086	26,363	132,088	2,540,500	5.20%
1984	98,166	7,985	24,762	130,913	2,606,200	5.02%
1985	97,148	9,270	28,913	135,331	2,626,900	5.15%
1986	110,044	10,271	31,728	152,043	2,699,700	5.63%
1987	116,635	11,254	33,307	161,196	2,728,200	5.91%
1988	117,015	11,656	37,992	166,663	2,762,800	6.03%
1989	118,428	12,293	40,393	171,114	2,752,800	6.22%
1990	127,395	12,684	40,410	180,489	2,748,100	6.57%
1991	119,469	11,916	50,745	182,130	2,804,100	6.50%
1992	131,402	12,909	50,549	194,860	2,792,300	6.98%
1993	132,814	14,664	53,616	201,094	2,856,400	7.04%
1994	138,293	16,460	60,662	215,415	2,929,000	7.35%
1995	152,102	17,433	57,451	226,986	3,000,700	7.56%
1996	155,898	18,888	64,028	238,814	3,160,800	7.56%
1997	168,457	19,131	72,534	260,122	3,234,800	8.04%
1998	155,906	19,119	73,411	248,436	3,276,100	7.58%
1999	157,685	19,807	77,295	254,787	3,319,600	7.68%
2000	154,676	19,466	74,959	249,101	3,374,200	7.38%
2001	141,079	19,894	77,574	238,547	3,425,900	6.96%
2002	135,870	19,680	77,509	233,059	3,474,000	6.71%
2003	124,933	19,409	80,461	224,803	3,472,500	6.47%
2004	122,077	18,244	85,905	226,226	3,515,900	6.43%
2005	122,870	16,752	93,086	232,708	3,538,100	6.58%

Source: Census and Statistics Department, Hong Kong, 2007

(4) *Variety of job offer*—construction offers a great variety of career opportunities. People with many different talents and educational backgrounds like managers, clerical workers, architects, engineers, quantity surveyors, truck drivers, trades workers, and construction helpers (Bureau of Labor Statistic, U.S. Department of Labor 2005c, Chow *et al.*, 2005).

(5) *Fragmentation*—In the UK in 1995, for instance, there were 163,000 registered construction companies, most employing fewer than eight people (Orange *et al.* 2000). This fragmentation has arisen due to the number of stakeholders and participants in the construction process from project inception through to project completion and beyond, each with divergent roles, goals, expertise and skills. It has resulted in unfavorable outcomes (Rinker 2000).

(6) *Competitive tendering*—Competitive bidding is the most commonly used method to award contracts. Each project is thus treated as a stand-alone transaction. Such approach is tempting for contractors to cut corners to recoup any loss that may arise. As such the quality of the work is often compromised (Rinker 2000).

(7) Products are frequently delivered by "project-based temporary multiple organizations" ("TMOs"), which exist only for duration of a single project (Cherns and Bryant 1984).

(8) Within a given project team different knowledge and skills are required at different times and for differing periods across the project duration, consequently only a small proportion of the project team remains in position for the complete duration of the project.

(9) Labor turnover within the industry is high. The use of short-term contracts is the norm.

(10) Construction has evolved from a very localized industry. Mega projects delivered by joint ventures composed of multi-national team members are now common (Chow *et al.*, 2005).

(11) There is significant seasonality in labor demand and poor labor practices exist in many of the smaller contracting companies.

(12) Adversarial and confrontational attitude is the norm rather than exception.

(13) Hazard wastes as residual product from construction—hazard wastes that might be generated during construction include ignitable paint wastes, other

ignitable wastes containing paint and varnish removers, paint brush cleaners, epoxy resins and adhesives, spent solvents, wastes containing toxic chemicals, and strong acid/alkaline wastes (EnviroSense 1996).

The TMO characteristic coupled with adversarial relationships has led to mistrust among project team members. This is one of the major problems faced by the construction industry. Barlow *et al.* (1997) nicely sums up this problem: "information and knowledge hiding are the two major barriers to learning lessons from projects that could lead to higher quality and productivity in future projects". These intrinsic characteristics are causing great concern to the construction participants as well as the clients. Some of the more immediate concerns are (CIRC 2001, Egan 1998):

(1) Built products are seldom defect-free
(2) Construction costs are comparatively high
(3) Costly and lengthy disputes
(4) Delay in product delivery
(5) Dissatisfied clients
(6) Insufficient investment in research and development and training
(7) Lack of skilled and trained workers
(8) Low productivity
(9) Poor communication

1.3 Need for Reform

In a survey of construction clients, Gallup (1995) discovered that substantial number of the projects which were the subject of their survey were delivered both late and over budget. A similar survey carried out by Morledge *et al.* (1996) found that of 215 commercial and industrial projects surveyed two thirds were delivered late. A report that focused on the problems of small and occasional clients of the industry (The Construction Clients' Forum 1997), found that some 60% of construction clients experienced problems when dealing with the construction industry. There is an urgent need to deal with these problems. For example, to reduce the fragmentation of the industry, greater collaboration and improving relations between participants in construction projects are encouraged (Orange *et al.* 2000). Several industry-wide reviews conducted in U.K. and Hong Kong have suggested reforms to that effect.

1.3.1 Latham Report—Constructing the Team

In 1994, one of the most influential reports on the review of the construction industry of U.K. was published. The study was called "The Government/Industry Review of Procurement and Contractual Arrangements in the U.K. Construction Industry". The terms of reference for the review include, inter alia, to consider current procurement and contractual arrangement with particular regard to methods of procurement, the organizational and management of the construction process, specific considerations include the desirability of a fair balance between the interests of, and the risks borne by, the client and the various parties involved in a project. The study aimed at making recommendations to Government, the construction industry and its client regarding reform to reduce conflict and litigation and encourage the industry's productivity and competitiveness. The final report was entitled "Constructing the Team", it served as a wake-up call against many problems faced by the construction industry that were identified as "ineffective", "adversarial", "fragmented" and "incapable of delivering for its clients". More than just ringing alarm bells, the Latham report laid the foundation for reform and gave the industry targets. The Report recommended that contracts should be founded upon principles of fairness, mutual trust and teamwork, rather than mistrust and confrontation. It was seen by the practitioners as a turning point for the construction industry, radically transforming relationships between clients and contractors. The report triggered the establishment of a number of taskforce groups in many parts of the world with the common view of improving the quality of construction industry. For example, the Construction Industry Board in the UK and the Construction Industry Institute in Hong Kong have been set up to oversee reform to the industry. In relation to construction contracting, Recommendation 19 of the report states that "specific advice should be given to public authorities so that they can experiment with partnering arrangements where appropriate long-term relationships can be built up. But the partner must initially be sought through a competitive tendering process, and for a specific period of time. Any partnering arrangement shall include mutually agreed and measurable targets for productivity improvements".

1.3.2 Egan Report—Rethinking Construction

"Rethinking Construction", the report of the Construction Task Force to the Deputy Prime Minister on the scope for improving the quality and efficiency of U.K. Construction, was published in 1998. The Task Force was established against a background of deep concern that the construction industry is under-achieving.

The Terms of Reference of the Task Force include: to advise the Deputy Prime Minister from the clients' perspective on the opportunities to improve efficiency and quality of delivery of U.K. Construction, to reinforce the impetus for change and to make the industry more responsive to customer needs.

The Task Force identified that fragmentation is one of the major problems in procurement practice. In this respect, partnering, a form of co-operative contracting, was seen as a tool to tackle fragmentation.

In the report (Egan 1998), partnering and examples of success are described as follows:

> "Partnering involves two or more organizations working together to improve performance through agreeing mutual objectives, devising a way for resolving dispute, committing themselves to continuous improvement, measuring progress and sharing the gains. In the Reading Construction Forum's best practice guides to partnering, 'Trusting the Team' and 'Seven Pillars of Partnering', it was suggested that where partnering is used over a series of construction projects 30% savings are common, and that a 50% reduction in cost and an 80% reduction in time are possible in some cases.
>
> Tesco Stores have reduced the capital cost of their stores since 1991, through partnering with a smaller supplier base with whom they have established long term relationship.

Argent, a major commercial developer, has used partnering arrangements to reduce the capital cost of its offices by 33% and total project time in some instances by 50% since 1991. They partner with three contractors and a limited number of specialist sub-contractors, consultants and designers."

Sir John Egan stated that effective projects would require collaborative efforts from all project participants, in which partnering played an important role; problems are to be resolved collaboratively by the entire team, not shoved off onto those least able to cope with them (Egan 1998). In his report, Egan (1998) identified five key drivers of change which set the agenda for the construction industry: committed leadership; a focus on the customer; integrated processes and teams; a quality driven agenda; and commitment to people. Under integrated processes and teams, the need for more integrated project team was emphasized: "the most successful enterprises do not fragment their operations—they work back from the customer's needs and focus on the product and the value it delivers to the customers. The process and production team are then integrated to deliver value to the customer efficiently and eliminate waste in all its form".

1.3.3 CIRC Report—"Construct for Excellent"

Notwithstanding the importance of the construction industry in contributing to the economy and development of Hong Kong, no review had ever been conducted before the Construction Industry Review Committee was appointed in 2000 to identify ways to improve the efficiency and cost-effectiveness of Hong Kong construction in terms of quality, customer satisfaction, timeliness in delivery and value for money. In the report (CIRC 2001), a package of recommendations was put forward with an aim to develop a new culture that focuses on delivering quality and cost-effective built products. Some of the measures suggested are directed to transform the construction industry by:

   i) Fostering a quality culture;
   ii) Achieving value in construction procurement;
   iii) Developing an efficient, innovative and productive industry; and
   iv) Achieving value in construction procurement.

Chapter five of the report is devoted to address ways in achieving value in construction procurement. The primary aim of construction procurement is to make sure the built facilities meet the clients' requirements in terms of quality, functionality and performance. Five issues relating to cost effective construction procurement were examined and these are:

    i) Selection of contractors and consultants;

    ii) Effective risk management and equitable contracting arrangements;

    iii) Dispute resolution;

    iv) A partnering approach; and

    v) Incentives for the project team to achieve better value.

A common thread underlying these issues is the attainment of an integrated project team sharing common goals. This begins with the selection of team members. It is advocated that non-price factors, in particular, past performance should be an integral evaluation component. To foster the cooperation of the team member, the contracting arrangements should be equitable. Dispute should be resolved as soon as possible. This calls for an efficient dispute resolution process as well as a collaborative forum. Team spirit can further be capitalized if appropriate incentives scheme can be derived to reward the project team that is able to achieve better value.

1.3.4 Paradigm Shift in Contracting Culture

Fragmentation at project level means that project participants are not working in a collaborative manner. They are used to their own distinctive cultures and working practices and not willing to change. The recommendations in Latham, Egan and CIRC suggest that there is a need for a major paradigm shift in contracting culture and practices. It is further advocated that the construction industry can only achieve this by engendering a spirit of compromise and collaboration, i.e., from a culture predominated by adversarial relationships between project participants to one focused on mutual understanding and collaboration.

It is now widely accepted that an integrated approach to construction with emphasis on teamwork and organization trust is the way forward (CIRC 2001). Amongst other recommendations, a wide adoption of co-operative contracting is highly recommended as an innovation strategy to improve the construction industry. The following chapters will describe the various approaches in achieving the principles of co-operative contracting.

1.4 Summary

The construction industry is one of the major pillars of the Hong Kong economy. The need for an efficient construction industry is therefore self-evident. Adversarial contracting behavior has been identified as one of the major contributors to the unsatisfactory performance of the industry. Several industry wide reviews have called for a paradigm shift in contracting culture. It is believed that co-operation built on trust is the vital contracting sentiment that would alleviate the adversarial nature of construction contracting behavior.

References

Barlow, J., M. Cohen, A. Jashapara, and Y. Simpson. 1997. *Towards Positive Partnering.* Bristol: The Policy Press.

Bureau of Labor Statistic. U.S. Department of Labor. 2005. Construction: Nature of the Industry. Under *Career Guide to Industries.* http://www.bls.gov/oco/cg/cgs003.htm#nature.

Bureau of Labor Statistic. U.S. Department of Labor. 2005. Construction: Working Conditions. Under *Career Guide to Industries.* http://www.bls.gov/oco/cg/cgs003.htm#nature.

Bureau of Labor Statistic, U.S. Department of Labor (2005). Construction: Occupations in the Industry. Under *Career Guide to Industries.* http://www.bls.gov/oco/cg/cgs003.htm#nature.

Census and Statistics Department, Hong Kong. 2007. *Hong Kong Statistics.* http://www.censtatd.gov.hk/hong_kong_statistics/index.jsp.

The Construction Clients' Forum (CCF). 1997. *Solving the Problems of Small and Occasional Clients of the Construction Industry.* Final Report. DoE, London.

Cherns, A., and D. Bryant. 1984. Studying the client's role in construction management. *Construction Management and Economics* 2: 177–184.

Chow, L. J., D. Then, and M. Skitmore. 2005. Characteristics of teamwork in Singapore construction projects. *Journal of Construction Research* 6: 15–46.

Construction Industry Review Committee (CIRC). *Construct for Excellence—Report of the Construction Industry Review Committee.* Hong Kong: HKSAR Government Print.

Dunican, P. 1985. UK construction industry, computer technology in construction. In *Proceedings of a Conference Organized by the Institution of Civil Engineers,* 1–12. London, UK.

Egan, J. 1998. *Rethinking Construction.* Department of the Environment, Transport and the Regions. London: HMSO.

EnviroSense. 1996. *Industry Overview of Construction.* http://es.epa.gov/techinfo/facts/construt.html.

Gallup Surveys. 1995. Customer survey. *Building* 28: 26–27.

Latham, S. M. 1994. *Constructing the Team: Final Report of the Government/Industry Review of Procurement and Contractual Arrangements in the UK Construction Industry.* London: HMSO.

Morledge, R., D. Bassett, and A. Sharif. 1996. Client time expectations and construction industry performance. In *RICS COBRA Conference Papers.* Bristol.

Orange, G., A. Burke, and J. Boam. 2000. Organizational learning in the UK construction industry: A knowledge management approach. In *School of Information Management—Research in Progress. Working Paper–IMRIP 2000–3.* http://www.lmu.ac.uk/ies/im/2000-3.pdf.

Rinker, M. E. 2000. *Nature of Construction Industry.* http://web.dcp.ufl.edu/katenah/bcn4773/NatureOfTheConstructionIndustry.htm.

2

Co-operative Contracting: In Search of Theoretical Anchors

Sai On Cheung

2.1 Introduction

The construction industry has for many years relied on formal contracts to define and enforce the obligations and rights of contracting parties. Conditions of contract have grown in size as project complexity increases. The changing roles amongst contracting parties within the construction supply chain call for, a new set of contractual arrangements designed to facilitate its operation is evidently required (Cook and Hancher 1990). A co-operative contracting paradigm appears to be the ideal fit. In this book, the set of contracting behavior that fosters co-operation among contracting parties is identified as co-operative contracting.

The obligations and rights of the contracting parties are typically stipulated in the contract. The orthodox approach assumes that all contracting parties are "rational maximizers"—always seek to maximize their own interests as much as possible. In this context, it is difficult to have wholehearted co-operation unless a supportive platform backed by an appropriately devised contractual framework is in place.

The commonly used design-bid-build procurement approach is adversarial in nature and therefore not amenable for co-operative contracting (Goddard 1997, Hancher 1989). There have been suggestions that contracting parties can attain mutual benefits through co-operation. By focusing on establishing a long-term relationship, adversarial attitude can be minimized (Provost and Lipscom 1989, Rubin and Lawson 1988). The notion of fostering co-operation in construction projects has attracted great interest. Notably, the use of partnering in construction projects is a good example of co-operative contracting (Hauck *et al.* 2004, Wong and Cheung 2005). Moreover co-operation is difficult to define, thus making this a contractual requirement can be problematic. In this chapter, the underpinnings of a number of contract theories will first be discussed. These will be compared with the empirically observed contracting behavior. The divergence so noted suggests the need for a more pragmatic contracting framework that embraces the concept of co-operation.

14

2.2 Contracting Behavior: Theory and Practice— An Overview

2.2.1 Classical Contract Theory

Some legal scholars (Macneil 1985, Macaulay 1985) classified contracts into two main categories: traditional and relational contracts. Traditional contracts are further divided into classical and neoclassical contracts (Macneil 1974). Figure 2.1 shows the different types of contract and their relationships.

Figure 2.1 Types of Contract

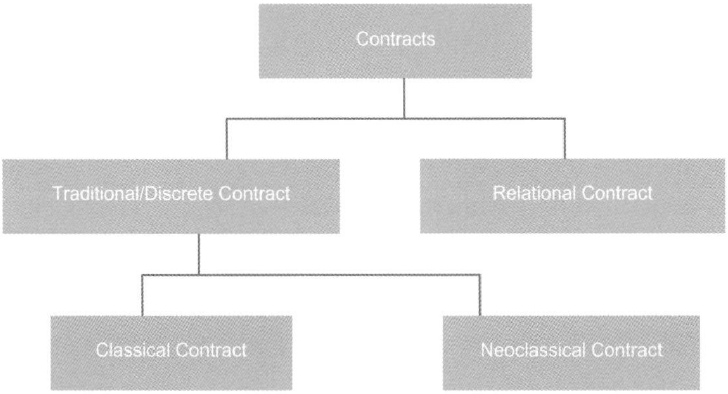

Classical model conceives contract as discrete exchange freely made by utility maximizers. With this regulatory framework, contracting parties in construction, would exhibit a self-interested contracting behavior, even to the point of taking advantage of the ignorance or vulnerability of the other party (Brownsword 1996). Contractors are also viewed as ruthless utility maximizers, exploiting every opportunity to advance their own self-interest. Brownsword (1996) concludes that the classical model treats contract as discrete rather than relational. In such context, classical contract law attempts to enhance discreteness and intensify "presentation", meaning "to make or render present in place or time; to cause to be perceived or realized at present" (Macneil,

1974). Complete presentiation portrays perfect contract whereby all relevant future contingencies pertaining to the contract are planned. A situation described as contingent claims contracting in the field of economics.

As the sophistication of constructed facilities grows, the development process has become progressively complex and of long duration. Not all future contingencies can be anticipated at the outset. Likewise, adjustments appropriate to the contingencies that may arise cannot be predetermined, in particular when impacts arising from contextual aspects have to be taken into account. The notion of discreteness therefore is inappropriate. Instead, contracting relation resembles those of a minisociety with a vast array of norms beyond those centered on the exchange and its immediate processes (Macneil 1978). Neoclassical contract law acknowledges that complete presentiation is prohibitively costly if not impossible, in particular, for long-term contracts.

In sum, a classical contract strives to cover as many contingencies as possible in order to reduce the possibility of claims and disputes. However, it is difficult to assess the extent and amount of contingencies when the market is volatile. The neoclassical contract regime is similar to that of the classical contract, but is "considerably modified in some, although by no means all, of its detail" (Macneil 1974). It involves a third party who assists in resolving claims or disputes and evaluates the performance of the two parties (Macneil 1978). The characteristics of classical and neoclassical contracts as described by Macneil (1978, 1987) and Lyons and Mehta (1998) are summarized in Table 2.1.

Table 2.1   Characteristics of Classical and Neoclassical Contracts

	Classical Contracts	Neoclassical Contracts
1.	It is used when the contract period is short.	The neoclassical contract is used with a specific fixed duration or task to be completed.
2.	Higher restrictions on personal interaction.	Personal interaction is relevant under this contract type.
3.	It is used when the transaction is "one-off" only and there will never be a future connection.	It is used when future cooperation opportunity exists.
4.	It allows a higher degree of discreteness and presentiation.	It allows a lower degree of discreteness and presentiation.

2.2.2 Critiques on Classical and Neoclassical Contract Law

According to Feinman (2000), classical contract law was the realm of consensual relations. The classical approach involved the application of relatively clear rules of legal doctrine, typically framed at a high level of generality and presenting dichotomous choices. The scope and method served the substance; within the realm of consensual relations, contract law simply developed ground rules for self-maximizing private ordering. Two types of criticism were raised. Firstly, internal criticism compared the ostensible rules with the results of cases and found that the rules did not explain the cases. No formal, general rules could. Secondly, external criticism stems from mapping these rules with actual contracting practices and argues that the Law's approach needed to be changed in order to serve the objectives of contract law. The external criticism is the result of contextualization, i.e., the more classical contract law was placed in context, the less sense it made.

Hillman (1997) provided a nice summary: "Neoclassical contract law is broadly the current status of contract law and apparently addresses the shortcomings of classical law rather than offering a different conception. The scope of neoclassical law is residual; no attempt is made to encompass all consensual transactions. It is fragmented, as the body of principles (the rules of formation, validation, performance, and remedies) is not necessarily applied in the same way in all types of case. The substantive core of neoclassical law is based on the assumption that parties act out of self-interest within a context of trade custom balanced by social values."

The critical weakness of the classical contract law is paucity of presentation. The fact of life is that not all risks can be presentiated, a scenario that is only achievable in spot contract. This fact thereby renders contracts incomplete. In construction, with the widely use of standard form of contract that had been comprehensively drafted, paucity of presentation may not be readily recognized. Nonetheless can paucity of presentation be reconciled as a conscious act of leaving certain risk allocation undecided? It may be true in some cases but it may just be the absence of a cost benefit or the risks will be allocated where they fall in fact. More interestingly, the absence of full presentation can be viewed as the contracting parties are having faith in each other. In this context, adjustments for unforeseen eventualities can be negotiated if so happens. In particular, where it is remote in having self-help remedies and cost of exit is prohibiting, the willingness to relinquish the advantages of presentation is a strong indicator of the faith in co-operative behavior. This sets the scene for co-operative contracting.

In his seminal paper on the optimal content of law school course, Macneil (1969) raised the question—Is there such a thing as contract? With his personal vested interest in economic, he reached the conclusion that contracts do exist. Nonetheless, his skepticism over the existence of "Law of Contracts" as it stood led him to search for other theoretical support—human behavior that is contractual. He suggested five common elements in contractual behavior in the following terms:

Co-operation
Every time a relationship seems properly to enjoy the label "contract", there is or has been some co-operation between or among the people connected with it.

Economic Exchange
Whenever behavior occurs which can be called contractual, there is always an element of economic exchange in it.

Planning for the Future
Contractual behavior involves mutual planning for the future, i.e., each party must decide that he wants what the other proposes to give and is willing to give what the other wants in return.

Potential External Sanctions
Potential sanctions external to the "contract" itself are incorporated or added to reinforce the relationship. These sanctions are not limited to be legal, instead non-legal sanctions such as business ostracism can often be more damaging.

Social Control and Manipulation

All contractual behavior is subject to social control and manipulation which may not take into account the interest and desire of those engaging in the behavior.

Macneil (1969) further explained how these five elements of contractual behavior present some significant challenges to the legal system:

> "Contracts are fundamentally mechanisms of co-operation, and only service conflict situations when things have gone wrong. As such the law largely has to deal with extreme cases. Two problems may arise:
>
> i) The law arising from pathological cases is not necessarily the optimum law for healthy cases; and
>
> ii) All cases need to fit into established principles and/or doctrines and the legal sanctions available are limited.

Moreover, where co-operation is required, the singular ingredient needed is an unspecific general willingness to co-operate whereby the relationship can be fostered. The fact that contract is a mechanism of economic exchange further amplifies the need to understand focus of exchange other than spot buy. Is there any conflict between co-operation and economic exchange because of the presumptions of individualism and maximizing behavior? It can perhaps be reconciled that exchange always involves a mutual goal of the parties, namely, the reciprocal transfer of values. The core of co-operation therefore can be exemplified if the exchange is extended over time. Whether he likes it or not a customer who enters a contract to purchase goods in the future enables the seller to plan his activities with a degree of assurance which he otherwise would probably have lacked. The customer is thereby co-operating in the seller's production or acquisition of the goods in a manner which would not have occurred had the customer simply purchased goods already produced." (Macneil 1969).

Building and engineering projects are typical exchange activities that require co-operation of the parties involved.

2.3 Contract Practices: When Relationship does Matter

In the case of [1] *Williams v Roffey Bros & Nicholls (Contractors) Ltd.*, a subcontractor approached the main contractor seeking for an increase in payment. The main contractor agreed to the request but later sought to render the payment increase unenforceable on the ground of lack of fresh consideration. However the Court of Appeal enforced the price increase on the ground that the main contractor had secured collateral commercial benefits, including changed working practices, avoidance of the penalty for delay, and avoidance of extra cost for the service of another subcontractor amount to fresh consideration (Collins 1996). From a contractual stand point, the main contractor had every right to reject the request of the subcontractor. Compensation is available to the main contractor should the subcontractor failed to honor his contractual obligations. However, the main contractor had agreed to the increase. This agreement was considered as an act derived from the self-interest of the main contractor. From a co-operative contracting perspective, the main contractor was aware that the subcontract had been under priced and the subcontractor was facing considerable loss. As such the behavior of the main contractor signaled a willingness to take into account the interests of the subcontractor for the sake of preserving business relation. Notwithstanding the contractual rights possessed by the main contractor as well as the expectation of a rational utility-maximizer, the agreement to an adjustment by the main contractor is thus commonly practiced (Beale and Dugdale, 1975). To reconcile this departure from anticipated behavior, Collins (1996) distinguishes the type of contract involved. With reference to Macneil's (1974) contract classification, i.e., rational utility-maximizing behavior characterized spot or discrete contracts and long-term contracts featuring co-operative efforts (Goldberg 1980). Collins (1996) suggests that some forms of catalyst are needed for the manifestation of co-operative behavior. Relationship is the key as it is unlikely that strangers to a contract would drive metamorphosis of rational self-interested contracting behavior.

---

[1] *Williams v Roffey Bros & Nicholls (Contractors) Ltd.* [1991] 1 QB1

Let us turn to some of the problems arising from long-term contracts involving complex contractual relationship raised by Collins (1996):

(1) The difficulties in specifying in advance the details of the expected contractual performance;

(2) The problem of coping with unexpected changes in circumstances, such as market conditions, during the period of performance of the contract;

(3) The problem of monitoring whether satisfactory contractual performance has in fact been provided; and

(4) The problem of guarding against opportunistic behavior like attempts to renegotiate terms of a contract.

Furthermore, Collin (1996) suggested four features representing distinctive behavioral patterns of long-term contracts:

(1) The rarity of legal enforcement of contractual rights;

(2) The paucity of express contractual allocation of risks;

(3) Elaborated planning against risks of opportunism; and

(4) The development of governance structures for the supervision of performance and the resolution of dispute.

In the context of co-operative contracting, the rarity of legal enforcement of contractual rights is of most interest. It may just be the simple fact that the cost involved in legal enforcement of contractual rights is too high. In construction, self-help remedies like security rights over property, insurance policies, performance bonds and indemnity are typical and provide reasonable safeguard against default. More stunningly, the loss of transaction specific investment such as skills, capital costs and research costs in long term contract deter breaking of the contractual relationship.

Nonetheless, the risk of opportunism is of real concern when one or both contracting parties have committed substantial assets for the contract. This transaction-specific investment creates the opportunity for forced renegotiation of contracts to the disadvantage of the party who has committed assets. The risk of opportunism is greater in long-term contracts which inevitably involve transaction-specific investments, a situation described by Collins (1996) as bilateral monopolies. Systems to monitor performance and to resolve disputes are therefore common and essential in long-term contracts. According to Williamson (1979) these governance structures can be viewed as a transaction costs economizing technique. As such, this should not be viewed as lack of a co-operative attitude. The most striking evidence of co-operative contracting behavior perhaps is thus the absence of a long-term contract even when there was the opportunity to make one (Collins 1996).

2.4 Towards a Co-operative Contracting Paradigm

As discussed in the preceding sections, construction contracting practices display noticeable departure from the assumptions based on which the classical contract law was developed. It can be said that the initial interest of co-operative contracting stems from the inadequacies of the classical contract law. As summarized by Campbell & Clay (1992),

> "The classical law assumes contracts to be the legal expression of the intentions of competent, self-interested parties who voluntarily enter into bargains on terms which they have understood and agreed at the time of acceptance. Still enshrined in the talismans of contract teaching—freedom of contract, sufficiency but not adequacy, consensus ad idem, etc.—this view is surely one of the most widely disseminated in higher education. But that there is something drastically wrong with it would, I think, be allowed by any sensitive commentator . . . "

Long-term contract makes most apparent the weaknesses of the classical law's assumptions that take discrete exchange to be the basic unit of contract analysis. The model of contract at the heart of the classical law is exemplified by a spot sale in a market which can be described as a place where "fareless buyers and sellers . . . meet . . . for an instant to exchange standardized goods at equilibrium prices" (Campbell and Clay 1992). Macneil (1974) advocates that in discrete transactions, the parties' commitment is also discrete with the assets involved readily transferable. The chance of under-compensation in remedies therefore is very low and supervision of performance by court is thus effective. Furthermore this type of spot exchange, will involve relatively little information that is not presentiated at the time of contract (Macneil 1974).

> "The discrete transaction is the perfect setting for maximizing recognition of exchange and its motivations. The narrow focus of the . . . transactional planning, the monetisation and measuring of what is exchanged, the minimum need for co-operation in planning and performance, the discreteness of the incidence of benefits and burdens, the specificity of obligation and the nature of potential sanctions, all go to guarantee the absence of room for anything but exchange and its motivations . . . Recognition by the parties of the prevalence and exclusivity of exchange motivations is inevitable in such . . . circumstances." (Macneil 1974).

Demsetz (1988) commented that this discrete domain is inadequate in governing complex construction contracts of long duration:

> " . . . in transactions of any complexity, it would be too costly to draw up contracts which would cover every contingency. Some aspects have to be left for interpretation when needed, and it is implicitly understood that it will be possible to agree on the meaning of the contract, even though one party loses."

The notion of presentiation therefore underpins the conception of classical contract law and is contradictory to the remedy and formation doctrines:

> "The modern commercial transaction is, in practice, apt to include provision for varying the terms of the exchange to suit the conditions applicable at the time of performance. Goods ordered for future delivery are likely to be supplied at prices ruling at the time of delivery; rise and fall clauses in building or construction works are the rule and not the exception; currency-variation clauses may well be included in international transactions . . . The rewards and penalties for guessing what the future will bring are no longer automatically thought of as being the natural consequences of success or failure in the skill and expertise of business activity."

According to Campbell & Clay (1992) there are three hypotheses about the nature of long-term contracting that had been empirically tested through the work of Macaulay (1966), Macneil (1974) and Williamson (1967):

(1) The documentation of long term contract tends to be open-ended, thus suggesting a rejection of the goal of presentation and in favor of explicit flexibility.

(2) In addition to the explicit sophistication of contract documents, there will be a co-operative recourse to extra-legal strategies to resolve problems which cannot be handled under these documents.

(3) The parties to the contract will, in all but the most extreme cases, adopt a co-operative rather than narrowly maximizing, strategic attitude to their own and the others' performance.

A symposium commemorating the immense contribution of Professor Macneil in the development of a relational contracting paradigm was held at Northwestern University School of Law in early 1999 (Feinman 2000). Although, the completeness and acceptance of a relational contract theory remains controversial; nonetheless the works of Macneil (1969, 1974, 1978) have provoked extensive discussion on the aforementioned inadequacies of classical and neoclassical contract law. Nonetheless, Macneil (1974) has not advocated a relational contract theory *per se*. Instead, he recalled, "I was simply exploring and trying to make sense of reality, the reality of what people are actually doing in the real-life world of exchange" (Macneil 1974). In this context, Macneil (2000) stated that the term "essential contract theory" can be used instead of relational contract theory. In essence it covers his descriptions of common contract behavior and norms, identified as:

(1) Role integrity (requiring consistency, involving internal conflict, and being inherently complex);
(2) Reciprocity (the principle of getting something back for something given);
(3) Implementation of planning;
(4) Effectuation of consent;
(5) Flexibility;

    (6)   Contractual solidarity;

    (7)   Restitution, reliance and expectation interests;

    (8)   Creation and restraint of power;

    (9)   Propriety of means, and

  (10)   Harmonization with the social matrix, that is with supracontract norms.

To summarize and quoting Feinman (2000), "the substance of relational contract theory is a refinement of neoclassical contract law. With relational contracts, greater attention is given to the desirability of fairness, co-operation and long-term interest. Thus, the substantive core of relational contract theory proceeds from two propositions: that a contract is fundamentally about co-operative social behavior and that contracts containing significant relational elements are the predominant form of contracting. This suggests that there is a baseline of obligation in contracting, one that arises out of the contract norm. This proposition is distinguished from the classical position that there is a baseline of no obligation, and from the neoclassical position that there is a core of self-interest affected at the periphery by custom and regulation. The precise content of the obligation is determined by the application of the relational method, i.e., the conformance to the common contract behavior and norms." A summary of the contract comparison proposed by Feinman (2000) is presented in Appendix 1 of this chapter (see page 45).

2.5 Co-operative Contracting in Context

Campbell and Harris (1993) opined that classical contract model is somewhat discredited due to its failing to give due weight to the idea of co-operation; a significant departure from actual behavior of contracting parties (Macaulay 1985).

Deakins and Wilkinson (1996) developed an institutional economic framework to explain how co-operation between economic agents is an essential element in the dynamic efficiency of a productive system, measured by its capacity to enhance information flows and engender a high rate of innovation in response to external change.

Mainstream economists have been advocating that competition and self-enforcement are means to ensure co-operation. Competition is derived from markets that provide the opportunities for individual agents to trade. If there are a good number of agents, each equally well informed, the market will operate freely to ensure co-operation and determine the relative prices of the factors of production. The critical prerequisite though is that each of the agents must be able to choose alternatives amongst a large number of substitutes thus commanding co-operation or suppressing opportunistic behavior. There are obvious deficiencies with this as free competition is an ideal rather than reality.

Deakins and Wilkinson (1996) advocate that there are at least two forms of co-operation: Inter and Intra. Division of labor within firm is a classic example of intra-firm co-operation. When this concept is extended to a supply chain, inter-firm co-operation is required. They further proposed that co-operation in co-operative contracting rests on the presence of, beyond the contract, organizational forms and institutional regulations that operate to a large degree independently of the motivations and behavioral traits of the economic actors themselves.

Brownsword (1996) suggests that a co-operative model of contract can derive support from at least three academic constituencies:

(1) Contract doctrines and adjudication of contract disputes is guided by co-operative ideals (Brownsword 1994);

(2) There are good empirical research findings that co-operative contracting is actually happening (Macaulay 1963, 1977, 1985, Beale and Dugdale 1975); and;

(3) Economic rationality often dictates co-operative conduct and that co-operative default rules for contract law may be welfare–maximizing (Trebilcock 1993).

Brownsword (1996) further expounds the concepts of co-operation in contract in the following terms:

(1) Co-operation happens more readily where there is joint investment in the contract with contracting parties who are mutually dependent on each other. Reciprocal co-operative performance is expected. Opportunistic behavior is eschewed.

(2) The incorporation of co-operation as a contract requirement is typically enshrined by the doctrine of good faith. Depending on the sophistication of the contracting parties, in particular, the extent of previous dealings and the institutional impact of the transaction, three levels of interpretation are possible:

    i) Contract specific interpretation dealing with performance and enforcement of the contract.

    ii) Relational interpretation is the co-operative ideal, the desired framework for all contractual dealings between the contracting parties.

    iii) Institutional interpretation goes beyond relational and the cooperative ideal applies to all dealings.

(3) The basis for co-operation prudential or moral? By prudential, it means such co-operative means are backed by ill self-interested motive. Moreover, for real co-operation that places long term interest over short term gains, the basis of co-operation should be regarded as a moral phenomenon. Prudence-based co-operation is not actually a case of co-operation.

(4) Utilitarian moral theory is relevant but not conclusive theoretical support for co-operation in contract.

Hence the concept of co-operation in contract, as suggested by Brownsword (1996) involves:

(1) Setting limits on the pursuit of self-interest by contractors;

(2) Invoking the idea of an overriding contractual community of interest; and

(3) Presupposing that co-operation ultimately rests on a moral foundation.

To put these in perspective, the meaning of contracting community of interest is important. The doctrine of good faith is the most direct analogy. Employing the comment (a) to section 205 of the Restatement (Second) of Contracts: "[good faith] excludes a variety of types of conduct characterized as involving 'bad faith' because they violate community standards of decency, fairness or reasonableness". According to Summers (1968) the concept of good faith is simply a reflection of so many unconnected instances of bad faith—the famous "Excluder" analysis of good faith.

Examples of bad faith behavior described by Summers (1968) are summarized in Table 2.2:

Table 2.2   Examples of Bad Faith Behavior (Adopted from Summers 1968)

Activities	Examples of Bad Faith Behavior
Negotiation	i) negotiating without serious intent to contract
	ii) abusing the privilege to withdraw a proposal or an offer
	iii) entering a deal not intending to perform
	iv) a seller's non-disclosure of known infirmities in goods
	v) taking advantage of another in driving a hand bargain
Performance of Contract	i) evasion of the spirit of the deal
	ii) lack of diligence and slacking off
	iii) willfully rendering only "substantial" performance
	iv) abuse of a power to specify contract terms
	v) abuse of a power to determine compliance
	vi) interference with or failing to co-operate in the other parties performance
Raising and Resolving Contract Disputes	i) conjuring up a dispute
	ii) adopting overreaching on "weaseling" interpretations and constructions of contract language
	iii) taking advantage of another to get a favorable readjustment or settlement of a dispute
Taking Remedial Action	i) abuse of a right to adequate assurances of future performance
	ii) wrongful refusal to accept the other's performance
	iii) willful failure to mitigate damages
	iv) abuse of a power to terminate

Brownsword (1996) further differentiates co-operative contracting with rational contracting which assumes contractors are rational and will aspire to co-operative behavior. This suggests that co-operation has to be backed by a rational law of contract.

Brownsword (1996) further elaborated three forms of rationality i) formal rational ii) instrumental rational and iii) moral rational:

> "Formal rationality is the fundamental but it has a logic problem. For example a liquidated damage clause is enforceable except when it is ruled as a penalty. Nonetheless, in most of these explicit contradictions, the exceptions are only used in case of emergency. Instrumental rationality operates at two levels; generic and specific. At the generic level, instrumental rationality requires law to be capable of guiding action. At the specific level, it concerns the appropriateness of particular legal interventions in given situations. Substantive moral rationality requires the supporting legal doctrine to be justifiable when set against rationally defensible moral criteria. Two forms of non-compliance are possible. Firstly, if the law is considered morally unacceptable to those the law is directed, non-compliance can be expected. Secondly, where a particular legal requirement is in conflict with the prudential interests of those to whom the law is specifically directed, non-compliance will occur."

To guard against non-compliance, two conventional regulatory truisms are noted: (i) Private Law remedial strategies rely on individual initiative, private law will fail to regulate effectively unless the prospective gains of litigation outweigh the costs; (ii) Disputants will only settle disputes through litigation if the present value of continuing relationships is low and the anticipated return from the litigation process is relatively high (Trubek 1975).

Notwithstanding the compelling arguments supporting the need for co-operative contracting behavior for complex projects, Macneil's (1974, 1978) relational contract theory is not widely accepted. Goetz and Scott (1981) describe that "a contract is relational to the extent that the parties are incapable of reducing important terms of the arrangement to well-defined obligations". Despite the seemingly lack of a firm legal footing, Goetz and Scott (1981) suggested that parties to a relational contract share the same objective of exploiting certain economics. Most importantly both parties shall come out as winner and share the benefits of the exchange. This is the key driver for co-operation which is the significant attribute to deal with contextual uncertainty and complexity that inhibit detailed specification on performance standard. Consequently more generic and general descriptions are used to define standard of performance required. Instead of relying on legal sanction, the ethical standard of the parties involved are called upon. "Best effort" clauses are probably the best that can be done to reduce this requirement to writing. The vagueness of this contractual governance renders control difficult if not impossible. Thus contractual devices such as incentivisation and sanction are included to foster performance. Sanctions such as liquidated damages and unilateral termination are regular in almost all construction contracts. However incentivisation, mostly derived from a "pain share, gain share" principle, is uncommon in construction contracting.

2.6 Bounded Rationality and Opportunism

Transaction cost economics aspires to describe "man as he is" (Coase 1984) in cognitive and self-interestedness respects. It works out of two key behavioral assumptions: bounded rationality and opportunism. The first of these implies that behavior is intendedly rational, but only limitedly so (Simon 1947), while opportunism anchors on the notion of self-interest seeking. According to Simon (1997), the term "bounded rationality" is used to designate rational choice that takes into account the cognitive limitations of the decision maker.

Williamson (1991) advocates two substantive approaches to business strategy: strategizing and economizing. It is further argued that economizing is more important among the two. Economizing involves effective adaptation and the elimination of waste. Williamson (1991) is in the view that the principal ramifications of these behavioral assumptions for economic organization therefore include:

(1) All complex contracts are unavoidably incomplete and many incentive alignment processes cannot be complemented (because of bounded rationality), thus most contingent adjustment mechanisms would fail for unanticipated eventualities.

(2) To rely on contract as promise is fraught with hazard (because of opportunism). Hence, Ideal forms of organization are disallowed.

(3) Added value will be realized by organizing in such a way as to economize on bounded rationality and to safeguard transactions against the hazards of opportunism. As such, transaction cost economizing is implicated.

According to Eisenberg (2001), a decision maker will only evaluate all possible options if the cost of searching for and possessing information were zero and human information possessing capabilities were perfect. In reality, such searches will normally be limited because of the cost required. Furthermore, decision makers are also bounded by their limitations on computational ability and on the ability to calculate consequences, understand implications, make comparative judgments on complex alternatives, organize and utilize memory (March 1978, Simon 1979). Thus, rationality is bounded by both the limitation on search for information and the computational power.

Simon (1972) advocated that theories that incorporate constraints on the information-processing capacities of the actor may be called theories of bounded rationality. There are at least three reasons against the notion of rationality in a real world situation:

(1) Uncertainty about the consequences;
(2) Incomplete information; and
(3) Complexity preventing the reasonable prediction.

The actual happenings as listed deviate from the presumption of having perfect information in a rational model; "Theories of bounded rationality are thus theories of decision making and choice that assume that the decision maker wishes to attain goals, and uses his or her mind as well as possible to that end; but theories that take into account in describing the decision process the actual capacities of the human mind" (Simon 1997). Furthermore, Simon (1957) introduced the concept of "satisfying" which suggests that "people will satisfy when they make a decision that satisfies and suffices for the purpose". Simon (1957) identified good administration as up keeping efficiency for which scare resources of an organization shall be deployed to accomplish its objective through rational behavior; "roughly speaking, rationality is concerned with the selection of preferred behavior alternatives in terms of some system of values whereby the consequences of behavior can be evaluated." Economic man as described by Simon (1957) is having a consistent system of preferences that guides him to choose among the options he has; he is aware of what these options are; there are no limit on the complexity of the analysis he can perform in order to select the best option; probability assessments are not intimidating. However, using the chess game as illustration, Simon (1972) suggested that instead of finding the optimal solution, choices are made when the decision maker regard an option is satisfactory because he can hardly be an economic man based on what he has in reality.

Dyner & Franco (2004) incorporated the concept of bounded rationality in modeling decision-making electricity users. It was found that these administrative men exhibit a kind of rational behavior that is compatible with the access to information and the computational capacities that are actually possessed by organizations, including man, in the kinds of environments in which such organizations exist (Simon 1957). The difference between "economic men" from "administrative men" is therefore significant as far as decision making is concerned. It is assumed that an economic man will evaluate all alternatives before making a choice. However sequential evaluation of options is the reality and the first satisfactory option is often chosen. The decision is therefore satisfying. The concept of bounded rationality is built on this apparently observable behavior.

Opportunism is another concept closely related to bounded rationality. Williamson (1993) suggested that "contractual incompleteness (due to bounded rationality) never give rise to contractual difficulties if parties to a contract can be relied on to self-enforce the agreement. As such, incompleteness, notwithstanding all gaps, omissions, errors etc. will be cured. A general clause such as "disclose all relevant information candidly and behave in a co-operative way during contract execution and at contract renewal intervals" may be included in the contract to formalize this desired state of operation. However, as Hobbes (1928) put it "Words . . . [are] too weak to hold men to the performance of their covenants". If opportunism is conceded to be the appropriate way to describe self-interest seeking, economic actors will break promises when it suits their purposes (Gauss 1952). In his seminal paper about contractual man, Williamson (1995) described that opportunism is the strong form of manifestation of self-interest behavior. He believes that individuals are opportunistic some of the time and that differential trust worthiness is rarely transparent before a contract is concluded. That explains why contemporary contracting practice places so much emphasis on selection and post-contract safeguard and control. These are organized to identify more disciplined contracting partners. However, this runs against the trusting spirits upon which the theory of co-operative contracting is built. Williamson (1985) further discusses the relationship between bounded rationality and opportunism. This view is presented in Table 2.3.

Table 2.3   Contracting Environment under Bounded Rationality and Opportunism

		Condition of Bounded Rationality	
		Absent	Admitted
Condition of Opportunism	Absent	Bliss[1]	"General clause" Contracting[2]
	Admitted	Comprehensive Contracting[3]	Serious Contracting Difficulties[4]

Source: Adopted from Williamson (1985)

1. An Utopia condition.
2. Example of a general clause like "I agreed candidly to disclose all relevant information and thereafter to propose and co-operate in joint profit-maximizing courses of action during the contract execution interval, the benefits of which gains will be divided without dispute according to the sharing ratio herein provided".
3. A scenario whereby prefect presentation is achieved.
4. Typical contracting environment in reality.

The significant implication from Table 2.3 is that conditions of construction contract are incomplete due to bounded rationally. If contracting parties behave as self-interest oriented economic actors, serious contracting difficulties can be expected. Section 2.7 discusses how trust can suppress opportunism in construction contracting.

2.7 Trust in Co-operative Contracting: Beyond Bounded Rationality and Opportunism

In Chapter one, the drivers for a paradigm shift in contracting culture has been expounded. This paradigm shift advocates co-operation among project participants and is intended to encourage clients, contractors, architects, engineers, surveyors, subcontractors, suppliers and end users to develop alternative project delivery systems that foster higher levels of co-operation throughout the project life cycle. The basic notion of "co-operation" implies seamless communication among parties as they work jointly towards a common goal. Traditional procurement systems have been widely criticised for their adversarial nature and the adoption of co-operative contracting for the procurement of construction facilities has become increasingly popular (Maxwell 2005).

Co-operation based contracting treasures relationship and the bonding among project participants are founded on the principles of equality, openness, problem-orientation, positive intent, empathy and has clear advantages over the confrontational approach (Hauck *et al.* 2004). For example, clients and their consultants will work more closely with contractors than would normally be exercised in traditional forms where the design and construction elements of a project are divorced. A co-operative approach to construction procurement can be regarded as the industry's affirmative response to the demands of the society (McDermott *et al.* 1994).

2.7.1 Contract as a Basis of Trust

Contract theory sees contract as a basis for trust since it limits the scope and incentives for opportunism (Woolthuis *et al.* 2005). Incomplete contract due to bounded rationality leaves rooms for interpretation whereby opportunism can creep in. Contracts that enhance trust can deter opportunism if the contracting parties behave trust-worthily. In this context, trust and contract are positively related, in fact, contract is a prerequisite for trust. It was further suggested that:

i) Trust will in general precede contracts;

ii) Trust and contracts can be both substitutes and complements;

iii) Relationships characterized by trust are most successful, this is in line with findings by social scientists that trust is an important condition to create an open and constructive atmosphere.

Lyons and Mehta (1997) analyze the role of trust in facilitating efficient exchange relation when parties to a contract are vulnerable to opportunistic behavior. They argued that trust can be derived from self-interest and/or through social orientation. Self-interest trust is forward looking while socially oriented trust has its root in the past. Undoubtedly contractual relationship can be facilitated when parties trust each other. Furthermore, trust can:

i) Reduce the costs of specification, monitoring and guarding against opportunistic behavior;

ii) Encourage better investment decisions; and

iii) Ensure rapid and flexible responses to unforeseen events.

It is the presence of trust that enables the transformation of relation, thus allowing the use of flexible contractual arrangements expressed in "loose" terms—a situation beyond the anticipation of the classical contract theory. Lyons and Mehta (1996) opine that trust is a meaningful concept for the parties to a social relation if and only if at least one party is exposed to an element of behavioral risk. Behavioral risk is experienced by one party when he or she is not sure about the behavior of the other who may take advantage of his or her good faith behavior. In other words the other is being opportunistic. This situation arises where the parties' regard to trust are asymmetric. This brings us to the issue of risk.

According to Sako (1992), contractual and goodwill risks are two distinct types of behavioral risk which are ameliorated by the presence of trust. Contractual risk is the risk that an agreement will be violated. Typically contractual violations include withholding payment, delay in completion. Trust will ameliorate the contracting environment as someone worthy of contractual trust is truthful and honest both ex ante and ex post with regard to behavioral pertaining to the agreement (Lyons and Mehta 1997).

Goodwill risk is the risk that a contracting party will act in his or her own narrow interest, and against the joint interests of both parties, should unanticipated event arise (Lyons and Mehta 1997). With an incomplete contract, the exposed party is facing the uncertainty of whether the other contracting party will respond flexibly and co-operatively to a request to modify the original agreement in some way. Furthermore, the other contracting party may even instigate opportunistic actions. The presence of trust facilitates the contracting parties to refrain from unfair advantage-taking. Elster (1989) advocates that goodwill trust presupposes contractual trust—"people may feel bound by the agreement that they would have reached had the unanticipated development been foreseen." What are then the bases of trust?

Weber (1922) described four ideal types of orientation of behavior in the following terms:

"i) Instrumentally–rational is a behavior that is determined by the expectations of the actor after taking purposeful calculation on the end, the means and the secondary results. This type of behavior is well recognized by mainstream economics.

ii) Value–rational behavior is the type that is for its own sake, thus involves self-conscious formulation for which the actor feels bound to obey. The loyalty, sense of duty and/or honour underpins value-rational behavior regardless of the cost to the actors.

iii) Affectual behavior has an emotional base. Accordingly, such behavior fulfils a need like revenge, and is independent of the ends it achieves, typically with no attempt to justify the behavior displayed.

iv) Most everyday behavior is traditional. At institutional level, the well practiced routines may be just part of a firm's corporate culture. In this regard, the reason for such behavior may just that it has always been done this way."

Lyons and Mehta (1997) advocate that there are two separable mechanisms that support trust: self-interest trust (SIT) and socially-oriented trust (SOT). Trust supported by instrumentally rational behavior is SIT. This form of trust is a direct response to the presence of behavioral risk with a focus on payoffs at the end. SIT therefore is forward looking, with contracting parties being trusting or trustworthy as long as positive return in the future can be derived. However, sociologists and anthropologists seem to be more interested in the ways in which individuals are bound together. Their views on behavior are thus located within a social arena that is norm based. As such the trust so derived is called socially-oriented trust (SOT). SOT is the product of an affectual, a traditional or a value-rational behavioral orientation as suggested by Lyons and Mehta (1997). SOT can be unstable and subject to exploitation by contracting party who displays instrumental-rational behavior. The sustainability of SOT relies heavily on recognized sanction and reward identifiable as industrial wide practices. What happened in previous dealings as well as during the course of the contract is therefore critical. Lyons and Mehta described that SOT has its roots firmly in the past.

Zucker (1986) provides another perspective on the origins of trust:

i)   Process-based trust that arises from a history of trustworthy interactions;

ii) Characteristic-based trust that rests on identifiable attributes such as family or religion, which are associated with trustworthy behavior; and

iii) Institution-based trust that is tied to institutions such as professions and bureaucracies.

While Process-based and characteristic-based trusts are clearly SOT, institution-based trust is more akin to SIT.

More recently, Nooteboom (2002) suggests two forces of trust: competence and intentional. Trust in the technical, cognitive, organizational, and communication competences of a partner is competence based. Trust that is related to the intentions of a partner towards the relationship, particularly in refraining from opportunism is intentional. Opportunism can have a passive/weak and an active/strong form. Lack of dedication in giving the best is an example of passive/weak form of opportunism. According to Williamson (1975), active form of opportunism entails "interest seeking with guile". Examples include lying, stealing and cheating to expropriate advantage from a partner.

Notwithstanding that isolated exchange is one-off, the parties most likely have some social relation in the past, be it from friendship, professional ethics or common membership of a business, religious or other community (Lyons and Mehta 1997). This suggests the presence of SOT even for isolated exchange. SOT in this respect is much more personal and identity is critical. In these contexts, detailed written contract documentation can be counter-productive because:

i) Narrow interpretation of the written clause may not truly reflect the spirit with which the contracting practices interact;

ii) Reducing to contract erodes the avenue to provide "convenience"—a highly effective tool to reinforce SOT; and

iii) Contemplating failure too explicitly casts doubt over the trustworthiness of the parties.

Even repeated dealings, if there is a definite number of repeats, are similar to one off exchange, in particular the penultimate one. Whereas when "infinite" repeats is anticipated, even self-interested contracting parties will adopt strategies that resemble SOT, at least until near the very end of the game. They remain co-operative because of the interests lying ahead. Although not truly SOT, opportunistic behavior is restrained.

Lyons and Mehta (1997) summarize their theory on trust as follows:

"i)   With SIT, the prime purpose of a contract is to use the force of the legal system to limit the potential for the other party to act opportunistically.

ii)   With SIT, no concession to the original agreement would be made without negotiation of a new gain.

iii)   With SOT, there is no need to specify detailed contingencies or penalties; but written documents act as a record of what has been agreed, and may even attempt to capture the spirit of the agreement.

iv)   With SOT, one of the parties may be willing to share the burden of some exogenous change in circumstances that would otherwise severely harm the other, even when that harm would not in itself compromise the continuance of the relationship."

2.8 Trust in Co-operative Contracting

Effecting co-operation in construction has been identified as one of the means to achieve the mission to revolutionize the construction industry. Matthews and Howell (2005) suggested the following maxims:

(1) Integrity and trust are essential for true collaborations;

(2) The long run is more important than the short run;

(3) Teams make better choices than individuals;

(4) In building a team, pre-qualify firms and select the right people;

(5) True creativity focuses on option generation, not just the selection of the "best idea";

(6) Change is inevitable: be prepared for it; and

(7) The basis for decision making should be facts and reasons, not opinions and emotions.

These maxims have been successfully demonstrated in a number of case studies including the Australian National Museum Project (Walker *et al.* 2000), MTRC TKE Rail Extension Project (Bayliss *et al.* 2004), OUC North Plant (Matthews and Howell 2005) and Welsh Water Treatment Works Project (Maxwell 2005).

Studies in the UK suggested that conventional project delivery system and its adversarial approach caused inefficiency and ineffectiveness, and recommended that co-operation could address these issues and had the potential for saving up to 30% over five years (Latham 1994). Similar studies also pointed that relationship based contracting could potentially achieve savings up to 30% (Bayliss *et al.* 2004). Some other benefits include (Matthews and Howell 2005):

(1) *Better communication and decision-making*—the project team members in particular will have more opportunities to voice out their concerns and express their views. This can develop a more integrated project team. With improved trust and communication among project members, each member is committed not only to perform his or her part but also assist the others when in need. This also helps improve project progress and the over quality of work.

(2) *Increasing productivity and innovation*—with project team members being more committed to contribute to the common project goals and objectives, less conflicting goals can be expected. This will foster harmonious working relationships, which in turns improve built product productivity and innovation.

(3) *Non-adversarial relationships*—problems arising during the duration of a project are resolved at appropriate levels with open and honest communication and mutual respect, before they become full-blown disputes that need to be dealt with by senior management or legal means. As a result, confrontation is generally replaced with co-operation and disputes are dealt with by non-adversarial means such as negotiation and mediation rather than litigation.

2.9 Summary

Like any system, successful co-operative contracting as a project delivery approach would not be possible without a well-defined framework. This chapter first examines the evolution of Classical Contract Law to Neoclassical Contract Law, and then followed by a discussion on the emergence of Relational Contract Law enunciated by Macneil (1974, 1978). Relational Contract Law has been identified as the theoretical anchor for co-operative contracting. Moreover, the presence of incomplete contract due to the impractical expectation of perfect presentation renders the risk of opportunism genuine in the practice of co-operative contracting. This chapter also introduces the roles of trust to counter this tension. On the practical side, the practice of co-operative contracting in construction can be identified in "project alliancing", "lean construction", "relational contracts", "integrated project system", "project team building", and "partnering" (Walker 2000). In fact, these terms are used interchangeably but share a common goal of ensuring mutual trust and respect among project team members. Notwithstanding the seemingly overwhelming enthusiasm in relationship based contracting, it is prudent to examine whether this conceptualization, to what extent can be applied to the different types of contracts within the construction process. Chapter three presents such a study.

References

Atiyah, P. S. 1979. *The Rise and Fall of Freedom of Contract.* Oxford: Clarendon Press.

Bayliss, R., S. O. Cheung, H. Suen, and S. P Wong. 2004. Effective partnering tools in construction: A case study on MTRC TKE contract 604 in Hong Kong. *International Journal of Project Management* 22: 253–263.

Beale, H., and T. Dugdale. 1975. Contracts between businessmen: Planning and the use of contractual remedies. *British Journal of Law and Society* 2: 45.

Brownsword, R. 1994. Two concepts of good faith. *Journal of Contract Law* 7: 197.

Brownsword, R. 1996. From co-operative contracting to a contract of co-operation. In *Contract and Economic Organization*, ed. Campbell and Vincent-Jones. Dartmouth Publishing Company.

Campbell, D., and Clay Suson. 1992. *Long-term Contracting: A Bibliography and Review of the Literature.* The Department of Law, City Polytechnic of Hong Kong.

Campbell, D., and D. Harris. 1993. Flexibility in long-term contractual relationships: The role of co-operation. *Journal of Law and Society* 20: 166.

Coase, R. H. 1984. The new institutional economics. *Journal of Institutional and Theoretical Economics* 140: 229–231.

Collins, H. 1996. Competing norms of contractual behavior. In *Contract and Economic Organization*, ed. Campbell and Vincent-Jones. Dartmouth Publishing Company.

Contract Journal. 2004. Article on Partnering. 28 February. 2001 Issues, 18–19.

Deakins, S., and F. Wilkinson. 1996. Contracts, co-operation and trust: The role of the institutional framework. In *Contracts and Economic Organization,* ed. Campbell and Vincent-Jones. Dartmouth Publishing Company.

Demsetz, H. 1988. *The Organization of Economic Activity.* Oxford: B. Blackwell.

Dyne, I., and C. J. Franco. 2004. Consumers bounded rationality: The case of competitive energy markets. *Systems Research and Behavioral Science* 21: 373–389.

Eisenberg, M. A. 2001. The theory of contracts. In *The Theory of Contract Law,* ed. Benson. Cambridge University Press.

Elster, J. 1989. *The Cement of Society.* Cambridge: Cambridge University Press.

Feinman, J. M. 2000. Relational contract theory in context. *Northwestern University Law Review* Vol 94(3): 737–748.

Gauss, C. 1952. *Niccolo Machiavelli, The Prince.* New York: New American Library.

Goetz, C. J., and R. E. Scott. 1981. Principles of relational contracts. *Virginia Law Review*, Vol 67(6): 1089–1150.

Hauck, A., D. Walker, K. Hampson, and R. Peters. 2004. Project alliancing at National Museum of Australia—collaborative process. *Journal of Construction Engineering and Management* (January/February): 143–152.

Hobbes, T. 1928. *Leviathan, or, the Matter, Forme and Power of Commonwealth, Ecclesiasticall and Civil (1651).* Oxford: Basil Blackwell.

Lyons, B., and J. Mehta. 1997. Contracts, opportunism and trust: Self-interest and social orientation. *Cambridge Journal of Economics* 21: 239–257.

Macaulay, S. 1963. Non-contractual relations in business. *American Sociological Review* 28: 55.

Macaulay, S. 1966. *Law and the Balance of Power: The Automobile Manufactures and Their Dealers.* Russell Sage Foundation.

Macaulay, S. 1985. An empirical view of contract. *Wisconsin Law Review.* 465.

Macaulay, S. 1977. Elegant models, empirical patterns and the complexities of contract. *Law and Society Review* 11: 507.

Macneil, I. R. 1969. Whither contracts? *Journal of Legal Education* 21: 403–418.

Macneil, I. R. 1974. The many futures of contracts. *Southern California Law Review* 47: 691–816.

Macneil, I. R. 1978. Contracts: Adjustments of long-term economic relations under classical, neoclassic and relational contract law. *Northwestern University Law Review* 47: 691–816.

Macneil, I. R. 2000. Relational contract theory: Challenges and queries. *Northwestern University Law Review* 94: 877–894.

March, J. G. 1978. Bounded rationality, ambiguity and the engineering of choice. *Bell Journal of Economics* 9: 587–608.

Matthews, O., and G. Howell. 2005. Integrated project delivery: An example of relational contracting. *Lean Construction Journal* 2(1): 46–61.

Maxwell, D. 2005. Alliancing—The future: a case study of a UK utility on its journey through partnering to alliancing. In *CII-HK Conference 2004 on Construction Partnering,* ed. A. Chan and D. Chan.

McDermott, P., T. Malaine, and D. Sheath. 1994. Construction procurement systems— What choice for the Third World? In *CIB 94*. http://cwis.livjm.ac.uk/blt/assets/ CIB94.pdf.

Nooteboom, B. 2002. *Trust: Forms, Foundations, Functions, Failures and Figure.* Cheltenham: Edward Elgar.

Sako, M. 1992. *Prices, Quality and Trust.* Cambridge: Cambridge University Press.

Simon, H. A. 1947. *Administrative Behavior.* New York: Macmillan.

Simon, H. A. 1957. *Models of Man, Social and Rational.* Wiley, New York, NY.

Simon, H. A. 1972. Theories of bounded rationality. In *Decision and Organization—A Volume in Honour of Jacob Marschak*, ed. C. B. McGuine & Roy Radner. North-Holland Publishing Company, Amsterdam-London.

Simon, H. A. 1979. Rational decision-making in business organizations. *American Economic Review* 69(4): 493–513.

Simon, H. A. 1997. Bounded rationality. In *Models of Bounded Rationality.* Vol 3. 291–293.

Summers, R. S. 1968. Good faith in general contract law and the sales provisions of the uniform commercial code. *Virginal Law Review* 54: 195.

Trebilcock, M. 1993. *The Limits of Freedom of Contract.* Cambridge: Harvard University Press.

Trubek, D. 1975. Notes on the comparative processes of handling disputes between economic enterprises. Paper presented at the US-Hungarian Conference on Contract Law and the Problems of Large Scale Economic Enterprise, New York.

Walker, D., K. Hampson, and R. Peters. 2000. Project alliancing and project partnering— What's the difference? Partner selection on the Australian National Museum Project —A case study. In CIBW92 Procurement System Symp. On *Information and Communication in Construction Procurement,* ed. A. Serpell, 641–655. Pontifica Universidad Catolica de Chile, Santiago, Chile.

Weber, M. 1922. *Economy and Society: An Outline of Interpretive Sociology.* 1978 ed. G. Roth and C. Wittich. Berkeley and London: University of California Press.

Williamson, O. E. 1967. The economics of defence contracting: Incentives and performance. In *Issues in Defence Economics,* ed. RN Mckean.

Williamson, O. E. 1975. *Markets and Hierarchies: Analysis and Antitrust Implications: A Study in the Economics of Internal Organization.* New York: Free Press.

Williamson, O. E. 1979. Transaction-cost economics: The governance of relations. *Journal of Law and Economics* 22: 33.

Williamson, O. E. 1985. The contractual man. In *The Economic Institutions of Capitalism,,* ed. Williamson.

Williamson, O. E. 1991. Strategizing, economizing, and economic organization. *Strategic Management Journal* Vol. 12: 75–94.

Williamson, O. E. 1993. Opportunism and its critics. *Managerial and Decision Economics* Vol. 14, 97–107.

Woolthuis, R. K., B. Hillebrand, and B. Nootboom. 2005. Trust, contract and relationship development. *Organization Studies* 26(6): 813–840.

Zucker, L. 1986. Production of trust: Institutional sources of economic structure 1840–1920. *Research in Organizational Behavior* Vol 8: 53–111.

Appendix 1

Compare and Contrast: Classical, Neoclassical and Relational Contract
(Adopted from Feinman 2000)

Issues		Classical and Neoclassical Contract Law	Relational Contract	A Critique of Relational Contract
Scope	What is the scope or structure of the field? What does it contain? How is its subject matter defined in relation to related fields?	Classical Contract Law assumes consensual relations Neoclassical Contract Law makes no attempt to encompass all consensual transaction	Relational Contracts can be governed by the core principles of contracts, as long as the courts applying the principles are sensitive to the factual differences in context.	Relational Contract Law simultaneously broadens and fragments the scope of contract law. The broadening in scope derives from the depict by Macneil (2000) that "Contract encompasses all human activities in which economic exchange is a significant factor." As such the scope of relational contract is very general. Neoclassical Contract Law allows factual differences whereby differential application of rules is accepted. Relational analysis argue for much more developed contextualization in distinguishing contracts according to more familiar categories.
Method	What method does the field use? What tools of analysis does it employ to decide cases?	Classical Contract Law involves the application of relatively clear rules of legal doctrine, typically framed at a high level of generality and presenting dichotomous choices. Under the Neoclassical paradigm, a mix of rules and standards is involved. Doctrines remain the integral part but of a much softer sort.	Relational method is also seen as consistent with neoclassical method BUT more vague standards are applied for ease of contextualization.	The method of Relational Contract Law is hard to describe. In addition to the contextual complexity, the factual background is also important. Nevertheless, both have to be understood within the contract norms like trade customs, rules of a trade association or professional organization.

Issues		Classical and Neoclassical Contract Law	Relational Contract	A Critique of Relational Contract
Substance	What is the field's substance? What are its core principles and purposes?	The ground rules of Classical Contract Law assume self-maximizing private ordering whereas under Neoclassical regime, parties act out of self-interest set within a context of trade custom and balanced by social values. **From Classical to Neoclassical** • Contextualization exposes the inadequacies of Classical Contract Law. • Internal criticism: the ostensible rules did not explain the cases. • External criticism: when the rules are situated in the world of actual contracting practice, it becomes apparent that the approach need to be changed to serve the objectives of Contract Law. The emerge of Neoclassical Contract Law was the responses to the internal and external criticisms. It is Neoclassical because it seeks to address the shortcomings of classical law rather than offering a wholly different conception of the law.	In relational contracts, greater attention needs to be paid to the desirability of fairness and cooperation; in relational contracts, short-term self-interest sometimes needs to be subordinated to long term self-interest.	The two propositions underpinning Relational Contract theory are: • Contract is fundamentally about cooperative social behavior, and • Contracts containing significant relational elements are the pre-dominant form of contracting. These propositions suggest a different baseline of obligation in Relational Contract as compared with classical and neoclassical contract law. In Classical Contract—the baseline is no obligation. In Neoclassical Contract—there is a core of self-interest affected at the periphery by custom and regulation. In Relational Contract—obligations to be determined by the application of Relational method.

3

Operationalizing Co-operation in Contracts

Sai On Cheung
Tak Wing Yiu

Acknowledgements
Part of the content of this chapter has been published in Volume 132(1) of the *Journal of Professional Issues in Engineering Education and Practice*, ASCE. The authors thank the permission of ASCE to publish the content therein in this chapter.

3.1 Introduction

In Chapter one, the contribution of the construction industry towards the economy of the Hong Kong has been outlined. A healthy construction industry is therefore no doubt important for Hong Kong. Nonetheless, many problems like claims, delays and inefficiency remain despite the call for reform has been so persistent. Working co-operatively within the construction supply chain is one of the key means to address these problems, in particular to quest for improved efficiency. The theory of relational contracting provides an instrumental support to the viability of co-operative working for relationship. Nonetheless the view on the legal basis of relational contracting, at least under the common law regime, remains divided.

The pragmatic view on contracting as expounded under the relational contracting domain has provided the much needed support for the structuring of innovative project delivery approach such as partnering. This chapter details a study that aimed to evaluate the degree of relationalism of the various forms of contracts commonly used during the construction process. As relational contracting theory advocates the importance of co-operation, the relational contracting framework is employed to gauge the readiness of construction contracts in operationalizing co-operation.

The study starts firstly with a relational analysis of construction contracts with the aim to highlight the transaction characteristics of relational contracts. Based on these characteristics, a Relational Index (RI) was introduced to compare the degree of relationalism of four types of construction contract; main contract, nominated subcontract, domestic subcontract and direct labor contract. The Relational Index is a measure of how aligned a particular type of contract is to the relational paradigm and hence is analogous to the use of a thermometer in measuring body temperature. In short, the higher the Relational Index, the more ready is the type of contract to operationalize co-operation.

3.2 Relational Analysis of Construction Contracts

One of the developments in the realization of a co-operative contract paradigm has been the furthering of fragment analysis on contract law (Fienman 2000). The law of franchise and employment are good examples. Furthermore, Fienman (2000) considered that commercial construction contracting that operates in a setting in which contracts, including forms of contract, are widely used by a mix of repeat and occasional players of different size and sophistication, in which interactions take place over time in a variety of settings, and in which problems always arise, is amenable to relational analysis.

According to Goetz and Scott (1981), "a contract is relational to the extent that the parties are incapable of reducing important terms of the arrangement to well-defined obligation". To this end, a relational contract is "incomplete", and hence some legally "ill-structured" provisions are often included, such as the obligation of one party ("the agent") to use its "best efforts" to carry on an activity beneficial to the principal and the concomitant right of the principal to terminate the relationship. The interpretation of these core provisions of relational contracts is often the prime source of costly litigation. Furthermore, the uncertainties over the legal treatment of these provisions impedes the ability of contracting parties to adjust to these special conditions that in turn induce relational contracting. Unlike Macneil's communitarian conceptualization of relational contracts, Goetz and Scott (1981) based theirs more on the economic opportunities accorded by relational contract. It was explained that each of the contracting parties wants a share of the benefits resulting from these economies and consequently seeks to structure the relationship so as to induce the other party to share the benefits of the exchange. The typical means to accomplish this is through specifying the performance standard of each party and then selecting a mechanism to ensure compliance with the agreed-upon standard.

Relational contracts are particularly suitable for projects filled with inherent complexity and uncertainty. As such, reducing performance standards to specific obligations is rather difficult as compared with conventional contracts. Thus, the parties would create unique, interdependent relationships, wherein unknown contingencies on the intricacy of the required responses may prevent the specification of precise performance standards. One notable example is the use of "best effort" clauses to articulate performance obligations in relational contracts. However, such an imprecise requirement calls for the use of control mechanisms like liquidated damages, incentives and unilateral termination provisions.

Construction contracts are conventionally fitted with detailed specifications that serve as performance standards. This works well as far as physical work and functionality are concerned. The rigidity and legal status that it carries firmly set the boundary of performance, a change of which invites conflict and dispute. The recent trend has been towards a wider use of the partnering approach and greater integration in finance, design, construction and operation as in Private Public Partnerships. Facilities development has therefore been undergoing major changes that require a fundamental revamp of contracting attitudes. In Macneil's terms (2000), the move has been from As-If-Discrete towards relational. Whereas construction has been just one of the phases in the whole development process, it is now more appropriate to view the complete development cycle as one project involving a number of key players: owner, designers, contractor, subcontractors and suppliers.

In these contexts, co-operative contracting is believed to represent a possible means of addressing the problems of adversarial relationships, mistrust and inefficient communication in the construction industry (Chan *et al.* 2004, Bayliss *et al.* 2004). In the last two decades, partnering as a form of co-operative contracting in construction has been used to deliver project with the aim of fostering a more collegial contracting environment. McInnis (2003a, 2003b) suggested that partnering contracts in construction exemplify relationalism, as partnering emphasizes relationship management. Reducing partnering behavior to explicit contractual requirements is not that straightforward. This is akin to the difficulty in specifying performance standards in relational contracts as suggested by Goetz and Scott (1981). In actual fact, a partnering agreement or charter is not even a formal contract; instead, it is treated as a moral contract (Barlow 2003). The underlying spirits of partnering, such as co-operation, trust, equality etc. are consonant with the concept of good faith (Heal 1999). However, the doctrine of good faith is a difficult concept to define (Colledge 1999). The applications of good faith in partnering will be discussed in further detail in Chapter eight.

In a common law system, there is no general obligation to observe good faith in the making or performing of a contract (O'Connor 1990). However, this does not mean that the courts allow unfair or unconscious acts in the formation or performance of a contract. Various rules or techniques, serving as substitutes of good faith, are adopted by the courts for achieving justice and fair results. The rationale behind this approach is that the common law system highly emphasizes the principle of freedom of contract and intervention in contract is done as an exception (Groves 1999).

The principle of freedom of contract will help in understanding the status of the doctrine of good faith in common law. The overriding principle of the freedom of contract can in fact be divided into two different but related forms. According to Cohen (1995), the first form is a positive one, which means that the parties are free to create a binding contract and make the terms of their agreement. The second form is a negative one meaning that the parties are free from obligations so long as a binding contract has not been concluded. It can be seen that the first positive freedom operates at the time of creation and performance of contract whereas the negative is relevant to the pre-contractual period. Summers (1968) employed an "excluder" theory that identifies good faith by way of contrast with the specific and variant forms of bad faith that the law prohibits (refer to Table 2.2 on P. 28).

Without a firm legal footing, the status of partnering agreement is at a crossroad. Notwithstanding this, Colledge (2000) presented a thorough analysis of the obligations of good faith in partnering in UK construction contracts and suggested that express provisions to use "best efforts", "best endeavors" or similar terms are akin to good faith, a relational feature identified by Goetz and Scott (1981). In actual fact, standard forms of contract for partnering projects are now commonly used, for instance, PPC 2000 (ACA 2000) and partnering option X12 of the New Engineering Contract (Telford 2005). PPC 2000 was reported a great success for project alliancing in the U.K. (Saunders and

Mosey 2005) and the New Engineering Contract has been identified as a suitable form for relational contracts (Gerrald 2005).

In sum, relational contracts in construction can be framed as being informal agreements involving an unwritten code of conduct that can powerfully bind the behavior between the contracting parties through features such as trust and relationship continuity (Baker *et al.* 2002, Deakin *et al.* 1994, Eisenberg 1995). A relational contract provides the means for sustaining long-term and complex contracts with a high degree of flexibility in order to allow parties to express their detailed knowledge in specific situations and adapt to new environments (Cheung 2002, Gundlach and Achrol 1993, Joskow 1987, 1990, Leffler and Ruker 1991, Macneil 1978, Macneil 1980, Swierczek 1994). The performance standard is governed by 'best effort' or "good faith" requirements. When there is a disagreement, a third party is involved in helping to match the contractual parties initially and a sanctioning party is called for if there is any breach of contract (Ellickson 1991, Grief 1993, Grief *et al.* 1994). The general characteristics of relational contracts are summarized in Table 3.1.

Table 3.1 Characteristics of Relational Contracts

Characteristics of Relational Contracts
1. The transaction is usually of a long duration.
2. Personal interaction is crucial.
3. The future co-operation opportunity is large.
4. There is a large degree of flexibility to cope with unforeseeable matters.
5. It is anti-discreteness and anti-presentiation.

Source: Cheung 2002, Josknow 1990, Leffler and Ruker 1991, Macneil 1987

3.2.1 How Relational are Construction Contracts?

Latham (1994) suggested that setting out effective terms and conditions, which include the contracting parties' duties in the construction process, is paramount in fostering a spirit of co-operation and teamwork. It is suggested that the choice of contract provision, to a certain extent, depends on the relationship of the contracting parties. An inappropriate choice of contract may not only affect the relationship between the contracting parties, but also the progress and the flow of interest, which may ultimately lead to construction disputes. As noted from the foregoing section, partnering in construction is akin to relational contracts notwithstanding the various implementation issues that have yet to be resolved. In Hong Kong, most partnering projects are still delivered using the traditional procurement framework. It is acknowledged that, because of the various types of contract within the construction process, the transaction characteristics are different and hence the "degree of relationalism" varies.

3.3 Development of a Relational Index (RI)

3.3.1 Critical Factors

There are a number of factors that would affect the choice of contract type. These factors are the building blocks for the construction of a Relational Index (RI), which gives an indication of how "relational" a contract type is. Reference is made to Section 3.2 under which factors influencing the choice of contract type are identified. Table 3.2 gives the eight selected factors for the study: co-operation, organizational culture, risk, trust, good faith, flexibility, use of alternative dispute resolution (ADR), and contract duration.

Co-operation
Hartnett (1990) described co-operation as a situation under which the contracting parties work towards the common goals and benefits of the project.

Organizational culture
Taking a broader perspective, organizational culture is the social energy which guides human behavior in an organization (Kilman *et al.* 1985). It provides implicit directions for the organization's members (Swierczek 1994).

Table 3.2   List of Factors for Relationship Measurement

Factors	References								
	1	2	3	4	5	6	7	8	9
1. Co-operation	√	√	√	√	√	√	√	√	√
2. Organizational Culture		√		√	√	√	√	√	
3. Risk	√				√	√	√	√	√
4. Trust		√	√		√	√	√	√	
5. Good faith		√	√		√	√	√	√	
6. Flexibility			√		√	√	√	√	
7. The Use of Alternative Dispute Resolution (ADR)			√			√		√	√
8. Contract Duration			√		√			√	√

Keys
1 — Cheung (2002)
2 — Feinman (1992)
3 — Goddard (1997)
4 — Halsbury (1973, 2000)
5 — Haugland (2003)
6 — Macedo Junior (1997)
7 — Macneil (1974b, 1975, 1978, 1987, 2001)
8 — Mcinnis (2003a, 2003b)
9 — Williamson (1979, 1985)

Risk

It refers to a situation in which the assessment of the probability of a certain event is statistically measurable. It relies upon the availability of known events for this purpose (Ashworth 1999).

Trust

The trust factor is a complex construct with multiple bases, levels and determinants (Hart 1988), and is often associated with situations involving personal conflict, uncertain outcomes and problem solving (Whitney 1996). It is also a prediction and expectation of future events. Varying in intensity, this is the confidence in and reliance upon the prediction (Rosenfeld *et al.* 1991).

Good Faith

It frames the contracting parties' behavior in acting honestly (Mcinnis 2003a, 2003b).

Flexibility

Bigsten *et al.* (1999) stated that flexibility arises when contractual performance is made explicitly or implicitly contingent upon external events affecting one of the parties, therefore making it a form of insurance and risk sharing. The riskier the environment, the higher the need for flexibility; and the higher the likely incidence of contract non-performance, the higher the expectation to renegotiate.

ADR

The use of ADR is an alternative to adjudicatory procedures. ADR includes conciliation, mediation, adjudication and the dispute resolution advisor system.

Contract Duration

It refers to the length of contract period. Generally, the longer the contract period is, the higher is the chance that changes will occur and thus a greater reliance on the relationship is needed to maintain the contractual bond.

3.3.2 Data Collection

Using the 8 factors as the basis for design of questionnaire, a survey was conducted in Hong Kong to solicit the views of practitioners on the degree of relationalism of different construction contracts. The respondents included those with extensive experience in construction contracts management. A list of contacts containing 80 potential respondents was created by searching the directories of professional bodies and companies' websites. Questionnaires were mailed to the prospective respondents. The respondents were asked to assess the degree of important of each of the 8 factors in affecting the continuing relationships between contracting parties in scale of 1 (least important) to 7 (most important).

3.4 Results and Discussion

A total of 48 completed questionnaires were received and that amounted to a 60% response rate. As to the professional background of the respondents, 28 of them were Quantity Surveyors, 20 were Project Managers/Engineers. They all have more than ten-year experience in contracts management. With the collected data, the ratings were analyzed by comparing the arithmetic mean scores for each factor. The overall results are summarized in Table 3.3.

Table 3.3   Arithmetic Mean Scores Comparison (Without Weightings)

Factors	Contract Types			
	Main Contract	Nominated Subcontract	Domestic Subcontract	Direct Labor Contract
Co-operation	5.50	4.89	5.39	4.72
Organizational Culture	4.50	4.38	5.08	4.58
Risk	5.15	4.58	5.08	4.35
Trust	5.55	4.91	5.27	4.59
Good Faith	4.96	4.58	5.23	4.65
Flexibility	4.74	4.35	4.74	4.35
The use of ADR	4.42	4.15	3.69	3.46
Contract Duration	4.81	4.77	4.69	4.00
Average Arithmetic Mean Score for a Particular Contract Type	4.95	4.58	4.90	4.34

The arithmetic means obtained in Table 3.3 were computed with equal weightings from the eight factors. This may not reflect the expected variations in contribution. As such, weightings were needed for each factor to show their relative importance. This was achieved by employing an analytical tool called the Analytical Hierarchy Process (AHP). It is a tool that can be used to determine the relative priorities of factors (Chua *et al.* 1999). The AHP process employs a pair-wise comparison matrix. Using the ExpertChoice™ software environment, the respondents were required to make judgments on the relative standings of the eight factors listed in the matrix table. Five procurement experts were invited to rank the factors under each one of the four contract types. A total of 20 pair-wise comparison matrices were performed. The pair-wise comparison matrix obtained from one of the respondents is given in Table 3.4.

Table 3.4   Pair-wise Comparison Matrix Obtained from One of the Respondents (For Main Contracts)

Factors/Criteria	Organizational Culture	Risk	Trust	Good Faith	Flexibility	Use of ADR	Contract Duration
Co-operation	5.0	3.0	1.0	3.0	1/5.0	5.0	3.0
Organizational Culture		1/3.0	1/5.0	1/3.0	1/5.0	1.0	1/3.0
Risk			1/5.0	1.0	1/5.0	3.0	3.0
Trust				5.0	1/3.0	5.0	5.0
Good Faith					1/5.0	3.0	3.0
Flexibility						5.0	5.0
Use of ADR							1.0

In Table 3.4, it can be seen that "co-operation" was compared with "organizational culture", followed by "risk" and then "trust" and so on. The pair-wise comparisons are guided by a nine-point scale as shown in Table 3.5. The procurement experts simply entered a scale in the empty cell corresponding to the factors being compared. The ExpertChoice™ software handles the pair-wise comparison and calculations automatically. The mathematics underlying the use of AHP to generate the relative importance ratings for the factors is based on linear algebra and graph theory. The relative standings are shown in descending order: Co-operation: 0.212, Flexibility: 0.180, Trust: 0.149, Risk: 0.134, Good Faith: 0.114, Organizational culture: 0.104, Contract Duration: 0.059 and Use of ADR: 0.048 (refer to Table 3.6).

The same assessment was repeated for the remaining contract types. At the end, with the assistance of the six procurement experts, five sets of importance weightings were obtained for each contract type.

Table 3.5   Nine-point Pair-wise Comparison Scale

Numerical Scale	Verbal Meaning	Explanation
1	Equal importance of both elements	Two elements contribute equally to the property
3	Moderate importance of one element over the other	Experience and judgment favor one element over the other
5	Strong importance of one element over the other	An element is strongly favored
7	Very strong importance of one element over the other	An element is very strongly dominant
9	Extreme importance of one element over the other	An element is favored at least an order of magnitude
2, 4, 6, 8	Intermediate values between the above adjacent values	Used for compromise between two judgments

Source: Saaty 1980

Table 3.6   Overall Importance Weightings and Arithmetic Mean Score Comparison (With Weightings)

Factor	Main Contract			Nominated Subcontract			Domestic Subcontract			Direct Labor Contract		
	Weighting (A)	Mean Score (B)	(A) x (B)	Weighting (A)	Mean Score (B)	(A) x (B)	Weighting (A)	Mean Score (B)	(A) x (B)	Weighting (A)	Mean Score (B)	(A) x (B)
Co-operation	0.212	5.50	1.17	0.190	4.89	0.93	0.131	5.39	0.71	0.140	4.72	0.66
Organizational Culture	0.104	4.50	0.47	0.104	4.38	0.46	0.105	5.08	0.53	0.129	4.58	0.59
Risk	0.134	5.15	0.69	0.090	4.58	0.41	0.125	5.08	0.64	0.134	4.35	0.58
Trust	0.149	5.55	0.83	0.164	4.91	0.81	0.155	5.27	0.82	0.214	4.59	0.98
Good Faith	0.114	4.96	0.57	0.111	4.58	0.51	0.153	5.23	0.80	0.152	4.65	0.71
Flexibility	0.180	4.74	0.85	0.198	4.35	0.86	0.214	4.74	1.01	0.129	4.35	0.56
The Use of ADR	0.048	4.42	0.21	0.049	4.15	0.20	0.047	3.69	0.17	0.043	3.46	0.15
Contract Duration	0.059	4.81	0.28	0.095	4.77	0.45	0.070	4.69	0.33	0.060	4.00	0.24
Average Arithmetic Mean Score (with weightings)	5.07			4.62			5.01			4.47		

In Table 3.6, the average arithmetic mean score for each of the factors were different depending on the contract type. To make it earlier to read, these scores are summarized in a Relational Index, which shows the relative position of each contract type.

Figure 3.1   Relational Index (RI)

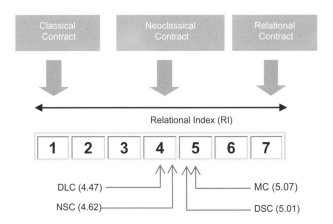

In Figure 3.1, it is clear that Main Contract type has the highest degree of relationalism, followed by Domestic Subcontract, Nominated Subcontract, and then Direct Labor Contract. The degree of relationalism can be assessed in two ways: 1) by comparing the numeric value of the relational index (Absolute Comparison); and 2) by comparing the relational indices in relative terms (Relative Comparison). In Figure 3.1, by absolute comparison, it shows that Neoclassical Contract was suggested for all of the contract types, a result that can be expected in a design-bid-build form of delivery. On the other hand, by relative comparison, it shows that Main Contracts and Domestic Subcontracts are more relational than Nominated Subcontracts and Direct Labor Contracts (refer to Table 3.7).

Table 3.7   Relationships between Contract Law Types and Commonly Used Contracts in Hong Kong

Contracts Commonly Used in Hong Kong Building Works	Relational Index (RI)	Suggestions	
		Absolute Comparison	Relative Comparison
Main Contract	5.07	Neoclassical Contract	Relational Contract
Domestic Subcontract	5.01	Neoclassical Contract	Relational Contract
Nominated Subcontract	4.62	Neoclassical Contract	Neoclassical Contract
Direct Labor Contract	4.47	Neoclassical Contract	Neoclassical Contract

3.5 Discussion

In actual fact, maintaining a good relationship between the client and main contractor is a prerequisite in fostering co-operation. This is also supported by the findings that "co-operation" is the most important factor for Main Contracts. Clients are the source of work and, therefore, maintaining a long-term business relationship with them is a means of survival from a main contractor's standpoint (Wong and Cheung 2004). This is more acute in Hong Kong as the construction market is dominated by a few mega-size developers. In practice, the main contractors and domestic subcontractors are partners and therefore majority of domestic subcontractors are very willing to cooperate with their main contractors. In many cases, domestic subcontractors even make changes to their own working habits simply to fit in their clients' (i.e., main contractors) working patterns. For this reason, Main Contract and Domestic Subcontract Types have similar relational indices. Nominated Subcontractors are very different to Domestic Subcontractors as they are specialist contractors appointed by clients. They have higher bargaining power because of their specialized skills. For this reason, the RI of Nominated Subcontracts is lower than that of Domestic Subcontracts. Turning to Domestic Labor Contracts, they provide "labor services" only. The issues involved in this kind of contract are relatively fewer and, in most cases, the contractual arrangements are straightforward. The degree of relationalism is therefore lower than the other contracts.

3.6 Summary

The high degree of fragmentation within the construction industry impedes the business relationship between the contracting parties. Establishing long-term relationships between the parties may improve this situation. The choice of contract types is the first step towards creating a platform for relationship development and management. Legal scholars have classified contract types into three types: classical; neoclassical; and relational contracts. This study seeks to examine the degree of relationalism of various contract types. This investigation is of value because notwithstanding the push for co-operation in contracting, it has to be recognized that different contract types exhibit transactional characteristics that may or may not be amenable to a relational paradigm. The degree of relationalism was measured in this study using a Relational Index framed by eight critical factors characterizing relationalism. These are Co-operation, Organizational Culture, Risk, Trust, Good Faith, Flexibility, Use of Alternative Dispute Resolution and Contract Duration. To enhance the discriminating power of the RI, five independent procurement experts assisted in determining weightings for the eight factors. The use of the Analytical Hierarchy Process (AHP) enhanced the reliability of the relative importance weightings as a result of its internal consistency checks.

The Relational Index suggests that, among the four contract types commonly featured in a traditional form of delivery, Main Contract and Domestic Subcontract types are more relational than Nominated Subcontract and Domestic Labor Contract types. These findings match well with general observations. The middle-range for the RI figures obtained (4.47 to 5.07 out of a total of 7) also suggests that the concepts of relational contracts may not be readily applicable to those contracts used in a design-bid-build form of delivery.

References

ACA. 2000. Standard Form of Contractor Project Partnering. ACA.

Ashworth, A. 1999. *Cost Studies of Building*. 3rd ed. England: Longman.

Baker, G., R. Gibbons, and K. J. Murphy. 2002. Relational contracts and the theory of the firm. *The Quarterly Journal of Economics* 117(1): 39–84.

Barlow, J. 2003. Partnering, lean production and the high performance work place. Working paper, School of Construction, Housing and Surveying, University of Westminster.

Bayliss, R., S. O. Cheung, C. H. Suen, and S. P. Wong. 2004. Effective partnering tools in construction: A case study on MTRC TKE contract 604 in Hong Kong. *International Journal of Project Management* 22(3): 253–263.

Bennett, J., and S. Jayes. 1998. *The Seven Pillars of Partnering: A Guide to Second Generation Partnering*. London: Thomas Telford.

Bennett, J., and S. Peace. 2002. *How to Use a Partnering Approach for a Construction Project: A Client Guide*. Englemere: The Chartered Institute of Building. London: CIOB.

Bigsten, A., P. Collier, S. Dercon, M. Fafchamps, B. Gauthier, J. W. Gunning, A. Oduro, R. Oostendrop, C. Patillo, M. Soderbom, F. Teal, and A. Zeufack. 1999. Contract flexibility and dispute resolution in African manufacturing. http://www.econ.ox.ac.uk.

Black, C., A. Akintoye, and E. Fitzgerald. 2000. An analysis of success factors and benefits of partnering in construction. *International Journal of Project Management* 18(6): 423–434.

Black, R. 2005. MTR partnering—A work in progress. CII-HK Conference 2004 on Construction Partnering, 9 December 2004, Hong Kong.

Chan, L. M., Kenneth T. W. Yiu, and S. O. Cheung. 2004. Re-thinking partnering in construction: Good faith and beyond. In *Proceedings of the World of Construction Project Management—1st International Conference, Toronto, Canada*. 484–493.

Cheung, S. O. 2002. Mapping dispute resolution mechanisms with construction contract types. *Cost Engineering* 44(8): 21–29.

Chua, D. K. H., Y. C. Kog, and P. K. Loh. 1999. Critical success factors for different project objectives. *Journal of Construction Engineering and Management* 125(3): 142–150.

Cohen, N. 1995. Pre-contractual duties: Two freedoms and the contract to negotiate. In *Good Faith and Fault in Contract Law,* ed. J. Beatson and D. Friedmann. Oxford: Clarendon Press.

Colledge, B. 1999. Good faith in construction contracts—The hidden agenda. *Construction Law Journal* 15(3): 288–299.

Colledge, B. 2000. Obligations of good faith in partnering in UK construction contracts. *The International Construction Law Review* 17(1): 175–201.

Construction Industry Board (CIB). 1997. *Partnering in the Team.* London: Thomas Telford.

Cook, E. L., and D. E. Hancher. 1990. Partnering: Contracting for the future. *Journal of Management in Engineering* 6(4): 431–446.

Deakin, S., C. Lane, and F. Wilkinson. 1994. Trust or law? Towards an integrated theory of contractual relations between firms. *Journal of Law and Society* 21(3): 329–349.

Eisenberg, M. A. 1995. Relational contracts. In *Good Faith and Fault in Contract Law*, ed. J. Beatson and D. Friedmann, 291–304. Oxford: Clarendon Press.

Ellickson, R. C. 1991. *Order without Law*. Cambridge: Harvard University Press.

Feinman, J. M. 1992. The last Promissory Estoppel article. *Fordham Law Review* 61: 303–316.

Feinman, J. M. 2000. Relational contract theory in context. *Northwestern University Law Review* 94: 737.

Gerrald, R. 2005. Relational contracts—NEC in perspective. *Lean Construction Journal* 2(1): 80–86.

Goddard, D. 1997. Long-term contracts: A law and economics perspective. *New Zealand Law Review* 4: 423–433.

Goetz, C. J., and R. E. Scott. 1981. Principles of relational contracts. *Virginia Law Review* 67(6): 1089–1150.

Grief, A. 1993. Contract enforceability and economic institutions in early trade: The Maghribi traders' coalition. *American Economic Review* 83(3): 525–548.

Grief, A., P. Milgrom and B. Weingast. 1994. Coordination, commitment, and enforcement: The case of the Merchant Guild. *Journal of Political Economy* 102(4): 745–776.

Gundlach, G. T., and R. S. Achrol. 1993. Governance in exchange—Contract law and its alternatives. *Journal of Public and Marketing* 12(2): 141–155.

Halsbury's Laws of England. 1973. 9(1). 4th ed. London: Butterworths.

Halsbury's Statutes of England and Wales. 2000. 4th ed. London: Butterworths.

Hancher, D. E. 1989. Partnering: Meeting the challenges of the future. *Interim Report of the Task Force on Partnering.* Texas: Construction Industry Institute. University of Texas.

Hart, K. M. 1988. A requisite for employee trust: Leadership. *Journal of Human Behavior* 25: 1–7.

Hartnett, J. T. 1990. Partnering. *The Military Engineer* 82(536): 20–21.

Hauck, A. J., D. H. T. Walker, K. D. Hampson, and R. J. Peters. 2004. Project alliancing at the National Museum of Australia—The collaborative process. *Journal of Construction Engineering and Management* 130(1): 143–153.

Haugland, S. A. 2003. Trust and relational contract. http://euro.nhh.no

Heal, A. J. 1999. Construction partnering: Good faith in theory and practice. *Construction Law Journal* 15(3): 288–299.

Hillman, R. 1997. The richness of contract law: An analysis and critique of contemporary theories of contract law. *The Michigan Law Review* 97(6): 173–90.

Joskow, P. L. 1987. Contract duration and transaction specific investment: Empirical evidence from the coal markets. *American Economic Review* 77(1): 168–186.

Joskow, P. L. 1990. The performance of long-term contracts: Further evidence from coal markets. *The Rand Journal of Economics* 21(2): 251–275.

Kilman, R., M. Saxton, and R. Serpa. 1985. *Gaining Control of the Corporate Culture.* San Francisco: Jossen–Bass.

Latham, M. 1994. *Constructing the Team: Final Report.* Joint Review of Procurement and Contractual Arrangements in the United Kingdom Construction Industry. HMSO, London.

Leffler, K. B., and R. R. Ruker. 1991. Transaction costs and the efficient organization of production: A study of timber-harvesting contracts. *The Journal of Political Economy* 99(5): 1060–1086.

Lyons, B., and J. J. Mehta. 1998. Private sector business contracts: The text between the lines. In *Contracts, Cooperation and Competition Studies in Economics, Management, and Law*, ed. S. Deakin and J. Michie, 43–66. New York: Oxford University Press.

Macaulay, S. 1985. An empirical view of contract. *Wisconsin Review*: 465.

Macedo, Junior R. P. 1997. Relational contract in Brazilian law. http://136.142.158.105/LASA97/portomacedopor.pdf

Macneil, I. R. 1969. Whither contracts? *Journal of Legal Education* 21: 403–418.

Macneil, I. R. 1974. The many futures of contracts. *Southern California Law Review* 47: 691–816.

Macneil, I. R. 1974b. Restatement (Second) of contracts and presentation. *Virginia Law Review* 60: 589–610.

Macneil, I. R. 1975. A primer of contract planning. *Southern California Law Review* 48: 627–704.

Macneil, I. R. 1978. Contracts: Adjustment of long-term economic relations under classical, neoclassical and relational contract law. *Northwestern University Law Review* 72(5): 854–905.

Macneil, I. R. 1985. Relational contract: What we do and do not know. *Wisconsin Law Review* 1985(3): 483–526.

Macneil, I. R. 1987. Relational contract theory as sociology: A reply to Professors Lindenberg and de Vos. *Journal of Institutional and Theoretical Economics* 143: 272–290.

Macneil, I. R. 2000. Relational contract theory: Challenges and queries. *Northwestern University Law Review*: 94(3), 877–907.

Macneil, I. R. 2001. *The Relational Theory of Contract: Selected Works of Ian Macneil.* London: Sweet & Maxwell.

Macncil, I. R. 1980. *The New Social Contract: An Inquiry into Modern Contractual Relations.* New Haven: Yale University Press.

McInnis, A. 2003a. The new engineering contract: Relational contracting, good faith and co-operation—Part I. *The International Construction Law Review* 20: 129–153.

McInnis, A. 2003b. The new engineering contract: Relational contracting, good faith and co-operation—Part II. *The International Construction Law Review* 20: 288–325.

O'Connor, J. F. 1990. *Good Faith in English Law.* Aldershot, Dartmouth Publishing Company Ltd.

Office of Government Commerce (OGC). 2003. The Integrated Project Team—Teamworking and Partnering. http://www.ogc.gov.uk/sdtoolkit/reference/ogc_library/achievingexcellence/ae5.pdf

Provost, R. D., and R. S. Lipscomb. 1989. Partnering: A case study. *Hydrocarbon Processing* 68(5): 46–51.

Rosenfeld, T., A. Warszawski, and A. Laufer. 1991. Quality circles in temporary organizations, lessons from construction projects. *International Journal of Project Management* 9(1): 21–28.

Rubin, D. K., and M. L. Lawson. 1988. Owners and engineers try long term partnering. *Engineering News-record* 220(19): 13–14.

Saaty, T. I. 1980. *The Analytical Hierarchy Process*. New York: McGraw-Hill Book Company.

Saunders, K., and D. Mosey. 2005. PPC 2000: Association of consultant architects standard form of project partnering contract. *Lean Construction Journal* 2(1): 62–66.

Schwartz, A. 1992. Relational contracts in the courts: An analysis of incomplete agreements and judicial strategies. *Journal of Legal Studies* 21(2): 271–318.

Summers, R. S. 1968. Good faith in general contract law and the sales provisions of the uniform commercial code. *Virginia Law Review* 54: 195.

Swierczek, F. W. 1994. Culture and conflict in joint ventures in Asia. *International Journal of Project Management* 12(1): 39–47.

Telford. 2005. *New Engineering Contract*. Thomas Telford Services Limited. Institution of Civil Engineers, U.K.

Uniform Commercial Code. 1968. Concept of good faith. *Virginia Law Review* 54: 195.

Walker, D. H. T., K. Hampson, and R. Peters. 2002. Project alliancing vs. project partnering: A case study of the Australian National Museum Project. *Supply Chain Management* 7(2): 83–91.

Whitney, J. O. 1996. *The Economics of Trust: Liberating Profits and Restoring Corporate Vitality*. New York: McGraw Hill.

Williamson, O. E. 1979. Transaction-cost economics: The governance of contractual relations. *The Journal of Law and Economics* 22: 233–261.

Williamson, O. E. 1985. *The Economic Institutions of Capitalism*. New York: The Free Press.

Wong, S. P., and S. O. Cheung 2004. Trust in construction partnering: The views of parties of a partnering dance. *The International Journal of Project Management* 22: 437–446.

Wong, S. P., and S. O. Cheung. 2005. Structural equation model of trust and partnering success. *Journal of Management in Engineering* 21(2): 70–80.

4

Partnering: An Exemplar of Co-operative Contracting in Construction

Sai On Cheung
Henry Suen

Acknowledgements
Part of the content of this chapter has been published in Volume 21 and 22 of the *International Journal of Project Management*. The authors thank the permission of Elsevier to publish the content therein in this chapter.

4.1 Introduction

The concept of partnering is no stranger to the construction industry nowadays. It has been advocated as a means to improve working relationships and project performance in terms of quality, cost, and time (Bayliss 2002, Bennett and Jayes 1995, CII 1991, Larson 1997). Successful applications of partnering in construction contracting have been reported in the UK, the United States, and Australia (CIIA 1996). In Japan, partnering is regarded as a normal way of working in the construction industry (Holti 1996). To successfully implement partnering, it has been suggested that the competence and skills of senior management are instrumental (Drexler and Larson 2000, Handy 1993). As pointed out by Bresnen and Marshal (2000), partnering has to be conceptualized as a management strategy that nurtures co-operative contracting with appropriate tools. While the collection of studies on partnering is rich, most of the studies focus on the description of types of partnering models, the management structures, and success factors. Li *et al.* (2000) categorized these studies under five main heads: empirical studies; non-empirical studies; relational; conceptual; and procedural. The literature and studies are largely prescriptive and anecdotal in nature. Where empirical studies do exist, they are mostly concerned with examining the partnering relationship (Kwan and Ofori 2001, Thompson 1998). The authors are of the view that, in addition to the existing descriptive models, research efforts should be directed towards the implementation issues and outcomes of the partnering approach. That view is shared by previous researchers in the field, such as Bresnen and Marshall (2000) and Li *et al.* (2000).

4.2 Definitions of Partnering

OGC (2003) summarizes partnering as: "an integrated project team working together to improve performance through agreeing mutual objectives, devising a way for resolving disputes and committing themselves to continuous improvement, measuring progress and sharing the gains." All the project team members have a shared goal of completing the project in a cost-effective and timely manner that is mutually beneficial. Partnering can either be a "one-off" for individual projects or a repeat process with the same team for a number of projects. Researchers in the field have come up with their definitions of what partnering is and these are summarized as follows:

The Associated General Contractors (1991)

"Partnering is a way of achieving an optimum relationship between a customer and a supplier. It is a method of doing business in which a person's word is his or her bond and where people accept responsibility for their actions . . . Partnering is not a business contract but recognition that every business contract includes an implied covenant of good faith".

Bennett and Jayes (1998)

"Partnering is a set of strategic actions which embody the mutual objectives of a number of firms achieved by co-operative decision-making aimed at using feedback to continuously improve their joint performance".

Construction Industry Institute of U.S.A. (1989)

"A long-term commitment between two or more organizations for the purposes of achieving specific business objectives by maximizing the effectiveness of each participant's resources. This requires changing traditional relationships to a shared culture without regard to organizational boundaries. The relationship is based on trust, dedication to common goals, and on an understanding of each other's individual expectations and values. Expected benefits include improved efficiency and cost effectiveness, increased opportunity for innovations, and the continuous improvements of quality products and services".

Cook and Hatcher (1991)

"All seek win-win solution, value is placed on the long-term relationships, trust and openness are norms, an environment for profit exists, all are encouraged to openly address any problem, all understand that neither benefits from exploitation of the other, innovation is encouraged and each partner is aware of the other's needs, concerns, objectives and is interested in helping their partner achieve such".

Crowley & Karim (1995)

"Partnering can be conceptually viewed as an organization that is formed by implementing a co-operative strategy that modifies and supplements the traditional boundaries between separate companies in a competitive market. Translating this concept to a working definition of project partnering, which is a method of transforming contractual relationships into a cohesive, project team that comply with a common set of goals and rely on clear procedures for resolving disputes in a timely and effective manner".

Hartnett (1990)

"The concept of partnering involves developing a co-operative management team with key players from the organizations involved in the construction contract. The team focuses on the common goals and benefits that are to be achieved through contract execution and develops processes to keep the team working towards them. This new approach to construction relationships highlights the common goals and relies on group dynamics to achieve them".

National Economic Development Council (NEDC) (1991)

"A contractual arrangement between a client and a chosen contractor which is either open-ended or has a term of a given number of years rather than the duration of a specific project. During the life of the arrangement, the contractor may be responsible for a number of projects, large or small and continuing maintenance work and shutdowns. The arrangement has either formal or informal mechanisms to promote co-operation between the parties.

4.3 Forms of Partnering

4.3.1 Project Partnering

It is partnering undertaken on a single project. At the end of the project, the partnering relationship is terminated and another relationship may commence on the next project. This kind of partnering is more likely to be adopted by public clients who have to ensure visible public accountability. It is often argued that project partnering is ineffective because trust and commitment could not be developed during a short contract period. It is asserted that the full benefits of partnering can be realized only through strategic partnering. It has been reported that project partnering can achieve savings of 2-10% in the total cost of construction due to improved communication between parties (OGC 2003).

4.3.2 Strategic Partnering

It is a long-term relationship that allows the two parties to work more effectively and efficiently. With this arrangement, the partners work together on series of projects and obtain full benefits from this arrangement as their relationship develops while trust and commonality of interests are fostered. It has been reported that strategic partnering can harvest significant savings of up to 30% in the cost of construction (OGC 2003, Black 2005).

4.3.3 Contractual and Non-contractual Partnering Agreement

Alternatively, another way of grouping is by examining the legal status of a partnering agreement. Non-contractual partnering agreement is a partnering agreement in which both parties do not intend the partnering charter to have any legal effect. On the other hand, a contractually binding partnering arrangement may be appropriate for either a project partnering or strategic partnering situation. In the project partnering situation, the purpose of creating a legal binding arrangement is to give legal effect to the issues in the partnering charter. While in the strategic partnering arrangement, the purpose even goes further, such as to provide a legal framework for regulating the long-term relationship between the parties (Butcher 1997). The continuum of partnering (in Figure 4.1) shows the anticipated performance of strategic, project and pseudo-partnering. According to Thompson and Sanders (1998), the most valued benefits of project and strategic partnering are reduction in project cost, reduction in total man hours and disputes. Pseudo-Partnering is of little value as it lacks the substances of genuine partnering such as common goals and co-operation.

Figure 4.1 Continuum of Partnering

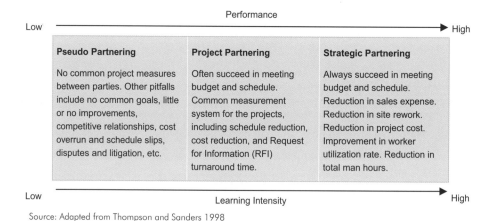

Source: Adapted from Thompson and Sanders 1998

Bennett and Peace (2002) pointed out that successful project and strategic partnering is founded on several key principles that help to define the partnering objectives and formulate measures for evaluation of partnering tools. These principles are:

    i) Involvement of key members of a project team;

    ii) Tender selection by quality, not lowest price;

    iii) Common processes such as shared IT;

   iv) A commitment to continuous improvement;

    v) Maintaining long-term business relationships; and

   vi) Fair risk-sharing.

4.4 Why Is It Worth Doing?

Numerous studies have identified the benefits of partnering in construction. Some of the most practical benefits include (Bennett and Jayes 1998, Black *et al.* 2000):

(1) Minimize the need for costly design changes

(2) Repeat good practice learned on earlier projects

(3) Minimize the risk of costly disputes—disputes are resolved promptly on site level

(4) Improvements in the quality of the construction

(5) Reductions in time and whole-life cost

(6) Integrate the whole supply chain—project members become more proactive and willing to contribute to common project goals

(7) With improved communication, less paper work

(8) Better understanding on clients' needs

The practical benefits of partnering can be further divided into three groups: benefits to project owner; benefits to project contractor; and benefits to both parties.

Benefits to Project Owner
Reduction of development cost:

- money spent in conflict resolution is reduced
- tendering cost is reduced
- saving due to sharing resources in partnering arrangement
- saving due to lower contingency sum in the price

Better quality product:

- greater understanding of client requirement by the contractors
- application of total quality management in partnering arrangement

Greater confidence on the availability of resources:

- by strategic partnering, the contractor agrees to work on a serial of projects

Benefits to Project Contractor

Lower overhead costs:

- due to less cost spent on bidding process
- effective communication with the client lowers the administrative cost

Steady flow of work:

- by strategic partnering, the client agrees to provide a series of projects

Greater profit:

- contract price will be negotiated by both parties rather than by competitive tender

Benefits to Both Parties

Reduction of litigation:

- due to mutual trust spirit of partnering
- better communication among the contracting parties
- pro-active dispute resolution mechanism in partnering arrangement

Reduction of contract period:

- due to reduced or eliminated tendering period
- the availability of structures for communication and co-operation
- reduced learning curves
- pro-active problem solving mechanism

Increased job safety:

- due to inclusion of safety goals and incentives in partnering agreement

Risks in Partnering

Like other management tools, partnering is not a panacea to all problems and the use of partnering does involve some risks. Some of the risks were identified by Black (1999) and summarized as follows:

- managers' unwillingness to relinquish control
- partners become complacent
- increasing dependence on the partner
- partners revert to adversarial approach
- absence of express provisions, an implied term that the contractor will proceed with reasonable diligence
- limited competition leading to cartels

4.5 The Evolution of a True Partnering

Using a simple relationship between two contracting parties as an example, the evolution of a true partnering can be divided into three key stages (Crowley and Karim 1995):

Stage 1
The two parties are in arms-length distance. At this stage, although a partnering arrangement has been established, the relationship remains vulnerable and not robust enough to avoid dispute.

Stage 2
The boundaries of the two parties are deformed for merging to take place. The "merged" boundary is still impermeable; some internal resources are re-organized by the individual parties and reserved as "collective resources". This is the beginning of the "formative partnering stage". This is when trust and share of information are emerged.

Stage 3
At this stage, the "merged" boundary is permeable for inter-organizational exchange to occur. Trust is enhanced and a partnering organization is formed permanently. The "merged" boundary becomes more permeable over time, resulting in the emergence of advanced features of partnering, such as implicit trust, long-term commitment, common goals, etc. (Thompson and Sanders 1998).

From an operational perspective, Li *et al.* (2000) developed a partnering process model which is intended to be used in most construction projects. They suggested that the partnering formation process should deal with the following issues:

i) The introduction of partnering to organizations, i.e., announcing the adoption of partnering as the blueprint for managing people and organizational culture within the project team.

ii) The identification of the needs for partnering, i.e., perform a thorough analysis to highlight any performance gap. Concerted efforts of all parties to bridge the gap shall be established.

iii) The selection of partnering companies, i.e., developing a list of partner selection criteria or measures.

iv) The organization of the partnering workshop, i.e., a charter for a project partnering.

v) The development of the partnering value during the workshop.

vi) The mobilization of the internal work process, i.e., developing goals of each party.

vii) The execution of the project, i.e., setting up teams to co-ordinate and monitor the partnering process.

Along a similar line of thinking of Li *et al.* (2000), Hellard (1995) suggests that a partnering process has six steps, starting from making clear intentions to the final evaluation of partnering process.

i) First step: Make clear intentions

ii) Second step: Plan for partnering workshop

iii) Third step: Designate the key persons to participate the workshop

iv) Fourth step: Conduct partnering workshop
   - □ creation of partnering charter
   - □ development of process for raising issues and resolving conflict
   - □ development of a joint evaluation process

v) Fifth step: Conduct follow-up workshop

vi) Sixth step: Final evaluations

The following gives a typical partnering process.

The partnering process is initiated by a partnering workshop/charter which occurs near the beginning of the project and includes team members from all firms involved in the project, i.e., the clients, contractors, subcontractors, architects and suppliers.

During this workshop, the individual companies set out project objectives and goals. This is referred to as the project charter, which set to bring out an important message to the team members-work together pro-actively! The charter should be reviewed during future partnering meetings to ensure the appropriateness of the mission and goals. During the workshop, all team members should define measurable yardsticks for the project. What is important is that they agree to measure team performance against stated goals vs. how one party is doing in relationship to others on the job. The measurement is against a "standard"—not against another part of the team. Win/Win relationships are essential if the partnering process is going to provide measurable results.

Another partnering tool developed during the kick-off workshop is the issue resolution process. Since partnering increases the rate and the number of issues identified, the team needs a tool to resolve issues efficiently (MBA 2005). The emphasis here is to establish a hierarchy on the project which encourages communication and creative problem solving as opposed to posturing and case building. Roles and responsibilities are defined and parallel points of communication are laid out. How issues are both identified and resolved are agreed to by the team. Senior management must reinforce in unequivocal terms and principles by which the team will conduct itself during the entire project. This commitment to "principle-centered leadership" is a key characteristic to successful partnering throughout the project.

The total team not only must use the tools on a daily basis, but they must be willing to trust their judgment and be accountable for their actions and team improvement. Monthly partnering meetings should be held to review the project charter and make necessary changes for improvements.

While the basic premise of working together in a co-operative team atmosphere is admirable, it is challenging in practice. Before rushing into partnering, all parties should understand that it is a process not an event. Like any process, if critical steps are disregarded, the chances for success are reduced. To ensure that the opportunities for a successfully partnered project are maximized, it is important that the following steps be considered and taken (MBA 2005):

Step 1: Prepare Well

The senior management of all parties should clearly understand what the partnering process involves, what their role needs to be, and to what they are committing their organizations. This can be accomplished by holding a preliminary meeting or conference call among senior-level executives before the actual retreat.

Step 2: Involve Senior Management

Senior management needs to be involved in the initial retreat as well as periodically meeting with the project team to ensure use of the partnering tools. An environment of operating fairly and with respect for the role of all the parties needs to be reinforced by senior management.

Step 3: Involve relevant partners

Without the proper people involved, partnering will fail. The client, main contractor, project manager, architect, subcontractors, suppliers and any other relevant parties need to be present at the partnering session and actively involved throughout future partnering meetings.

Step 4: Clearly Defined Partnering Team Leaders

Common sense indicated that the partnering will succeed or fail based on how well the project leaders work together and how well they reinforce the partnering concepts on a day-to-day basis in the field. These leaders need to identify potential problem areas and make partnering a part of the every-day language of all workers on the project.

Step 5: Follow-Up and Reinforce

The lack of formal follow-up is the primary cause of failed partnered projects. As a minimum, project team leaders from all parties involved should meet on a regular basis to discuss how the partnering is progressing and to measure success. Without formal follow-up, momentum may be lost and team members may fall back into old ways of doing business when difficult issues surface.

The partnering process will only produce positive results if project team members are willing to commit themselves to it throughout the life of the project. Anticipated outcome of partnering include productivity improvement, positive schedule impact, and reduction in claims. At best, these will create real dollar savings and new profit opportunities for the parties involved.

4.6 The Australian Experience

Australia is pursuing best practice through the use of partnering which is gaining increasing momentum not only in the building and construction industry but in other industries as well. A number of reports have identified the need for clients and contractors to better define the project and respective risks prior to commencement to avoid time and cost overrun and, most importantly, poor business relations between the parties (CIIA 1996). These reports identified the need for more co-operation and open dialogue between contracting parties. In practice, that means getting the client, the consultants, the main contractor, and the subcontractors of a project together to discuss important issues at an early stage so that they could identify major areas and better understand the total procurement picture. Master Builders Australia (MBA) in responding to these reports formed a special group which set to investigate and report on what partnering can offer to the industry. It was subsequently found that the US and the Australian construction industry were facing the same problems such as lack of trust and "blaming" culture. With an intention to change the culture, MBA pioneered the development of partnering in the construction industry by bringing in partnering experts to Australia in 1992 and 1993 to conduct a series of seminars and meet with industry leaders. The result is promising with a large number of construction projects adopting partnering since then. The exact number and value of projects is not fully known. According to a study reported by MBA in 2005, about 350 to 400 projects had used partnering, with a total contract value in exceed of $14 billion Australian dollars (MBA 2005). Some notable examples of these "partnered" projects include:

(1) The Garden City Shopping Centre;
(2) Nepean Hospital, NT University Tourism and Hospitality faculty;
(3) Hotham Redevelopment;
(4) South Road Connector East Arm Port Darwin; and
(5) Woodford Correctional Centre

MBA believes that the concept of good faith and fair dealing in contracting may be the missing link to underpin the co-operative attitude of the partners. The opportunities for the development of such strategic partnering processes as distinct from project partnering has greater potential in private sector than to public sector where public accountability is more strictly pursued (MBA 2005). Through its legal department and its strategic alliances with Australia's leading construction law firms, MBA provide services to assist contractors and clients to use partnering. MBA believes that "the contract establishes the legal relationship, partnering establishes the working relationship". As such, using a partnering approach is not a new way of doing business. As a matter of fact, many have been practicing co-operative contracting for quite some time. Partnering helps create an environment where trust and teamwork prevent disputes, foster a co-operative bond for everyone's benefit, and facilitate the completion of a successful project.

Partnering charter is often seen as a symbol of commitment. All stakeholders' interests are considered in creating mutual goals and there is commitment to satisfying each stakeholder's requirements for a successful project by utilizing win-win thinking. Successful implementation of partnering requires trust, mutual objectives and timely evaluation.

4.6.1 Trust

Through the development of personal relationships and communication about each stakeholder's risks and goals, trust is established among members of a project team.

4.6.2 Development of Mutual Goals

At a partnering workshop the stakeholders identify all respective goals for the project in which their interests overlap. These jointly-developed and mutually agreed goals may include achieving savings, meeting the financial goals of each party, limiting cost growth, limiting review periods for contract submittals, early completion, no lost time because of injuries, minimizing paperwork generated for the purpose of case building or posturing, no litigation, or other goals specific to the nature of the project.

4.6.3 Timely Evaluation

In order to ensure implementation, the stakeholders agree to a plan for periodic joint evaluation based on the mutually agreed goals. Timely communication and decision making not only saves money, but also can keep a problem from growing into a dispute. In the partnering workshop the stakeholders develop mechanisms for encouraging rapid issue resolution, including the escalation of unresolved issues to the next level of management.

4.7 The UK Experience

Partnering has become an extremely popular management tool in the UK construction industry in recent years. It has become widely used to increase the levels of client satisfaction—a mutual benefit for client and contractor. Yet the primary focus of partnering has been on the principles, from which "best practice" can be derived. This is because much of the drive towards partnering has come from clients. As a result, attention to the underlying concepts and theory has been neglected to a considerable degree. This lack of attention can be tolerated provided that there are real reasons to assume that partnering will work in the medium and long term. The main emphasis to date has been upon project partnering. In conceptual terms, partnering has mainly been viewed from a procurement perspective. This is a consequence of partnering being driven by the client. Coupled with the emphasis upon project partnering, the focus has primarily been upon tactical project objectives. However, strategic partnering is the form that can provide the greatest potential benefits for the contractor. The significant outcome of the current focus is that the benefits largely accrue to the client (Smyth 2002). The key benefits are:

(1) Cost reductions
(2) Provision of tailored service
(3) Client satisfaction

From an operational perspective, partnering has been viewed as a procurement issue. Bennett and Jayes (1998) produced some of the seminal work in the UK and commented that "The UK construction industry needs partnering in order to achieve tough targets set for it in the Latham Report." Therefore the focus is upon meeting client procurement objectives. However, it is not a simple procurement issue. Bennett and Jayes (1998) see the three key objectives of partnering as achieving mutual objectives, problem resolution and continuous improvement. They have broken mutual objectives down into a "management by objectives" type of approach, which is couched in terms of procurement: "Partnering is widely found to help customers get their project teams to focus on their needs and objectives."

The procurement theme continues through quality, cost efficiencies, speed and other largely project related issues. The contractor is required to act in a responsive way via the project team to the client. This is the foundation of how partnering is being practiced, the only benefit to the contractor being securing the sale. This may lead to an increase in market share. It will only lead to higher profitability where the premium margin outstrips the costs of meeting the increased demands of the client. In practice, this will seldom be realized where the contractor is competing on a project-by-project basis for partnering contracts.

In an attempt to understand the behavior of contracting parties under a partnering arrangement, Smyth (1999) made use of a matrix of buyer and seller relationships. The matrix locates procurement in its various forms. Smyth (1999) identified three main strategies for buying and selling:

(1) Competitive

(2) Co-operative

(3) Command

The competitive strategy is the classical market approach. If the client and contractor behave in a competitive way, then both are acting independently of each other resulting in a perfect market. This has been the preventing condition. The command strategy can be adopted where either the supplier or the client has a dominant position in the market. They can employ leverage from their positions to dictate the terms of the contract. In construction, this monopolizing position is quite common given the fact that, in many cases, clients usually have the upper hand in negotiating the terms of the contract. The co-operative strategy is one in which mutual advantage can be achieved. Relationship marketing, and its tool partnering, is to be seen as part of this. These three strategies exist on both sides of the exchange process—buying and selling.

Partnering is about moving away from open competition towards a more co-operative strategy. However, this does not mean that the co-operation is truly mutual. Partnering always requires a co-operative strategy on behalf of the contractor. The procurement emphasis means that the client is maintaining a commanding position for the following reasons:

(1) Adoption of a project bias to procurement

(2) Having an imbalance between the number of contracts and size of contracts in relation to the number of contractors being partnered

(3) Having intensive tender competition among partners for contracts

(4) Changing the ground rules of partnering

(5) Low levels of investment in the partnering relationship

(6) Being disloyal to contractors

4.7.1 Lesson from Gleeson (2005)

Glesson is a public-listed building and construction firm in the UK and is at the forefront of partnering initiative in the construction industry. It is estimated that over

65% of its workload is carried out on a partnering basis. The partnering ethos is fundamentally built into management practices and procedures. In Glesson (2005), partnering is described as:

i) Committed to our clients objectives

ii) Minimizing the contractual interfaces

iii) Joint problems, joint effort, joint solutions

iv) Allocation of risk

v) Efficient delivery and value for money

vi) Engaging the supply chain

vii) Teamwork, integration and co-location

The operating principles adopted by Glesson (2005) are as follows:

i) Foundation of a core team, ideally co-located

ii) Run facilitated workshops to develop understanding of each partner's point of view and objectives

iii) Focused on problem solving innovation and the sharing of Best Practice (rather than waste effort in adversarial approaches)

iv) Engaging the supply chain in the design process at the earliest possible stage—thus designing out many potential construction problems and costs

v) Extend partnering throughout the supply chain to minimize contractual interfaces and maximize benefits

vi) Establish trust through "open book" accounting and budget reviews

vii) Target cost saving incentives with shared rewards

viii) Agree a problem resolution process

ix) Use value management and value engineering to continuously manage risk and improve quality and efficiency

x) Measure performance on a regular basis in the drive for feedback and continuous improvement

4.8 An Overview of Partnering in the Hong Kong Construction Industry

The concept of "Partnering" was first introduced in the North District Hospital Project in 1994. Since then there has been a number of construction projects administrated under the Partnering arrangement. As Chan *et al.* (2002) summarized: "an increasing trend in project partnering in construction has been observed in the public, private and infrastructure sectors over the past decade, with a proven track record of success".

In the public sector, the spate of cases regarding defective piles found at a number of Housing Authority's sites made some headlines in the local newspapers in Hong Kong in 2000. This raised the public's concern over the quality of housing in general. In response, the Housing Authority launched a Quality Reform with a view to improve the built quality of public housing. A total of 50 initiatives were proposed in the Reform and "Building up a partnering framework" is one of them. In this regard, three pilot works projects were selected to try out the use of "Partnering". The result was encouraging. As Ms. Ada Fung, Assistant Director (Development and Procurement) of Housing Department, said (2004): "we are convinced that partnering as well as project partnering will help to build rapport, achieve better working relationship among all stakeholders and achieve better project results with better reliability . . . we are positive about partnering and we mean it".[1] C. S. Wai, Deputy Secretary of the Environment, Transport and Works Bureau (ETWB) in a paper entitled "adoption of partnering in public works projects[2]" mentioned that the ETWB has adopted a step-by-step approach to introducing partnering into its public works contracts. The Secretary went on to say that, since 2001, the ETWB has adopted non-contractual partnering in more than 30 public works contracts for building, civil engineering, electrical and mechanical works.

---

[1] Fung A. (2004) Project Partnering: The Housing Authority's Experience. A paper presented at the CII-HK Conference 2004 on Construction Partnering on December 9, 2004.

[2] Wai C. S. (2004) Adoption of Partnering in Public Works Projects. A paper presented at the CII-HK Conference 2004 on Construction Partnering on December 9, 2004.

In the private sector, the Partnering Charter implemented for "Charter House" project, which involved the demolition and redevelopment of the 35-year-old Swire House, was the first of its kind in that sector. The decision to adopt Partnering was made in late 1997, a time of serious economic concerns in Hong Kong. That decision turned out to be a wise one. The remarks of Mr. James Robinson, Executive Director (Projects) of Hong Kong Land Limited, say it all: "many months ago, Mike Arnold co-hosted the keynote speech at the launch of the Charter House Partnering Strategy. At the time, I shared with those attending the launch, my view of partnering as a means to facilitate the often-quoted but rarely achieved "win-win" outcome. Now, with Charter House's Occupation Permit safely tucked in our proverbial back-pocket and with the first of our tenants preparing to move in, I am pleased to reflect on our collective achievements". [3] HongKong Land moved on from the partnering success on Charter House and employed the same approach for the Landmark Refurbishment and Redevelopment Scheme.

In terms of partnering in infrastructure projects, the Drainage Services Department (DSD) embarked on its first partnering journey in early 2001—the same year the Construction Industry Board's report entitled "Construction for Excellence" was published. DSD's first experience of Partnering was described by Mr. Raymond Cheung, Director of Drainage Services Department, as "Bitter-Sweet". [4] It was a major stormwater drainage tunnel project formed a key component of DSD's overall drainage improvement strategy to solve the flooding problem in Mongkok. It was 'bitter', as explained by Mr. Cheung, because "the current standard form of contract for public works is incompatible with partnering as it provides no financial incentives for fostering a true partnering approach". Mr. Cheung went on to suggest the need of a risk-sharing mechanism for use in partnering projects. It was "sweet" because the non-contractual approach adopted for the drainage tunnel contract improved the

[3] Robinson J. (2004) Balance in Partnering: An Owner's Perspective. A paper presented at the CII-HK Conference 2004 on Construction Partnering on December 9, 2004.

[4] Cheung R. and Kan F. (2004) Bitter-Sweet Experience of Non Contractual Partnering on a DSD Contract. A paper presented to the CII-HK Conference 2004 on Construction Partnering on December 9, 2004.

communication and fostered the spirit of project teams. Turning from drainage works to tunnels, the New Territories East Development Office (NTEDO) had their first taste of Partnering in two projects in the New Territories: Contract No.ST89/02-Sha Tin Heights Tunnel and Contract No.ST79/02-Road T3 and associated roadworks. In both cases, the Partnering was arranged post contract award. Mr. John Climas, Deputy Project Manager for the projects, said[5] the major benefits, arising from the partnering arrangement, are completion on-time, under budget, with few significant claims and with high quality end products.

MTRC Corporation Ltd. (MTRC) is recognized as the forerunner in partnering in the last decade or so. The authors have had the privilege of assisting Mr. Roger Bayliss, Executive Project Manager of MTRC, in the assessment of the effects of Partnering in construction projects. MTRC has been using partnering with its contractors since 1998. Partnering has gone through stages of development; from initial non-contractual partnering to the introduction of elements of contractual incentivisation, and followed by recently target cost contracts. These all serve the same purpose—to foster a highly co-operative relationship between the client, the contractor and the consultants. The Tseung Kwan O Extension (TKE) Project (1998–2002) was an exemplar—it sets high standards for followers of partnering to follow. In the following section, the TKE and its use of partnering will be discussed in full. To recap and forecast the development of partnering in the Hong Kong construction industry, the authors borrow a passage of Mr. Russell Black (2004): "proactive partnering inevitably raises questions revolving around the moral hazard of 'helping the contractor', or similar terminology such as 'doing the contractor's job for him' and 'the contractor is taking advantage'. The issues are real and deserve on-going consideration on the job. There is no simple answer; the issue has to be taken within the context of the relationship. Senior management must have full visibility; but decision making should remain on site. The ethical probity of agreements should be debated openly and not evaded".

---

[5] Climas J. and Kam C. W. (2004) HKSAR Government's Experience of Partnering on Two Major Highway Projects. A paper presented to the CII-HK Conference 2004 on Construction Partnering on December 9, 2004.

4.9 A Case Study of MTRC TKE Contract 604

The Mass Transit Railway Corporation (MTRC) is one of the pioneers in the industry to implement partnering in the delivery of construction projects in Hong Kong. It has always been the MTRC Project Division's view that effective project management relies on co-operative working relationships between consultants, contractors and clients. It was in 1998, when the construction of the Tseng Kwan O Extension was about to commence, that the Project Division of MTRC recognized the need for a co-operative contracting environment in order to achieve the quality construction anticipated by the Corporation (Bayliss 2001). In view of the above, shortly before the construction phase of the TKE project, small working group was set up to investigate the concept of partnering. The working group was particularly interested in the benefits that partnering can offer and how partnering might be introduced to the Project Division's projects. After months of studies and field trips of investigation, the working group came to a conclusion that the introduction of partnering would improve cost effectiveness, give greater programme certainty, result in better communication, more co-operation and greater responsiveness to problems (Bayliss 2002). It was also stated that there was real potential for introducing partnering to the Project Division's project management toolkit. The working group suggested that partnering may not be appropriate for all construction projects, but should be introduced on a step-by-step basis. It was proposed that the relationship management (soft) side of partnering should be introduced initially, with the commercial (hard) issues to follow.

It was immediately after the award of the contract for TKE 604 that partnering was introduced. The TKE 604 contract is the 6th operational line of MTRC. It comprises two main routes. The first route is a diversion of the existing Kwun Tong Line from the existing Lam Tin Station, passing through Yau Tong and terminating at Tiu Keng Leng. The second route involves the construction of a new Tseung Kwan O Line, connecting with the existing Eastern Harbour Crossing on the Kowloon side, passing through Yau Tong and beyond into Tseung Kwan O New Town. The TKE project consisted of 13 civil contracts, 4 building services contracts and 17 E&M contracts. The TKE Contract

604 was one of the 13 civil contracts and involved the construction of Yau Tong Station, which involved site formation, foundations and structure, services and architectural works etc. The contract was awarded to a Japanese construction company, Kumagai Gumi (Japan) Ltd. (KG). Considerable efforts were directed in ensuring the success of the partnering endeavors These included the use of external consultants to facilitate partnering workshops, the development of a partnering charter, running monthly review meetings, organizing social functions, publishing a partnering newsletter, and the use of an Incentivisation scheme.

Partnering tools such as workshops, review meetings, incentives and the like are often used to effect the partnering spirit (Bayliss 2001, 2002). However, there are still considerable debates as to what form these partnering tools should take and how effective they are (Barlow et al. 1997). Hence, there is a need for an in-depth study of the various partnering tools and their relative effectiveness in implementing partnering. Empirical evidence to support the successful use or otherwise of the partnering tools would therefore be highly valuable to the construction community (Bennett and Jayes 1995, Bresnen and Marshall 2000). This paper reports a case study of MTRC TKE Contract 604 in which partnering was adopted to instill, foster and maintain partnering spirit.

The primary objective of this case study was to identify and analyze the partnering tools that had been used effective in effecting the intent of a partnering delivery. As partnering tools used in many partnering projects are similar, the findings in this case study provide a good indication of their potential effects. The objectives of the study were achieved through:

(1) Providing a detailed account of the partnering mechanism
(2) Developing an evaluation framework for the assessment of the effectiveness of the partnering tools
(3) Evaluating the effectiveness of the partnering tools as perceived by the users
(4) Collecting empirical support to the perceptive views in (3) above from contract data

Furthermore, the project data kept also provides the opportunity to explore the importance of behavior project personnel in partnering endeavors and this has been reported (Cheung *et al.* 2003).

4.9.1 An Account of the Partnering Efforts

To obtain a complete picture of how partnering was implemented in TKE Contract 604, interviews with the key contract participants (both MTRC staff and KG staff) were first conducted. The partnering tools deployed for TKE Contract 604 are discussed seriatim.

Executive Partnering Workshop

In practice, partnering is normally instigated through a "top-down" communication channel. Thus, to embrace the partnering concept within the MTR Corporation, an executive partnering workshop (facilitated by an external consultant) was conducted with the MTRC executives. The main function was to introduce the MTRC senior management to the new mode of project delivery. This set the scene for the implementation of partnering for the TKE project.

Contract Specific Partnering Workshop

A contract specific partnering workshop was held, involving the key contract participants and facilitated by an independent facilitator. Based on the contract characteristics, a contract specific partnering charter was agreed and signed by the contract participants. The charter defined the project mission, goals and objectives, and measures. Some of the objectives addressed in the TKE Contract 604 partnering charter included co-operative working relationships, timely completion, quality service, waste reduction, and effective dispute resolution. Subsequently the partnering concept was reinforced by follow-up workshops and junior workshops, the latter facilitated by MTRC staff.

Monthly Partnering Review Meetings

The partnering status of the contract was monitored in monthly partnering review meetings. To facilitate the function of the review meetings, a questionnaire was designed to assess the contract partnering status as perceived by the participants. The review meetings were held monthly and attended by both the MTRC and KG's staff. In essence, the contract partnering status was assessed by considering thirteen attributes. These include trust, honesty, communication, co-operation, programme, quality, safety, financial objectives, job satisfaction, resources, waste minimization, 3rd parties' needs, and dispute resolution. Before each review meeting, attendees were requested to fill out the questionnaire. A sample of the questionnaire and details of the review meetings can be found in the paper mentioned in the previous sections (Cheung *et al.* 2003). The score chart (see Appendix 1 on page 110) summarized the partnering scores and the averages against each attribute were compared with the scores recorded in the previous meetings. Being honest, open and self-critical was paramount in the score assignment process. During the meetings, the scores and their changes were discussed, a process through which problems or areas of concern were unveiled. The changes of scores would be interpreted as a reflection of the achievement of a certain partnering attribute for the period immediately before the review meeting. A decline suggested something negative might have happened and the attention of the management team was thereby drawn. It was also a deliberate activity to arrange the chairmanship of the meeting to be rotated among the meeting attendees. It is believed that such an arrangement encourages the parties' active participation and open communication, improved the commitment of participants and helped to move away from the hierarchical structure normally associated with construction projects.

Social Functions

Social functions, such as boat-trip, football competition, karaoke, Chinese "Yum-Cha" were organized to maintain team spirit, which is essential to effective and cooperative working environment. Interviews with senior management suggested that these functions help to develop interpersonal communication and improve trust.

Newsletters

A monthly partnering newsletter, called "Win-Win", was issued to the contract participants to report success stories of partnering. These included articles on partnering experience, pictures taken from informal functions, latest news and developments in partnering, etc. These served to raise staff awareness of the benefits of partnering, as well as to maintain the culture of partnering within the project team.

Incentivisation

An Incentivisation Agreement (IA) in the form of a supplemental agreement was signed by the parties about mid-way through the contract, after thorough discussion in the preceding few months. The Incentivisation Agreement identified the risk exposure to be borne solely by MTRC or KG. The remaining risks would be shared between MTRC and KG. A target cost was established for dealing with those shared risks and a pain share/gain share formula was agreed whereby savings or excesses of the target cost would be shared by MTRC and KG.

The use of the IA crystallized the risk elements and reduced uncertainty with regard to responsibility. The incentive for KG was his entitlement to a portion of the savings if the costs associated with the shared risks could be contained. In this respect, value management concept was applied to reduce costs, thus achieving a "Win-Win" for both MTRC and KG.

4.9.2 An Evaluation Framework

The second phase of the case study involved the development of an evaluative framework for the partnering mechanism. There were two aspects to be evaluated: firstly, the effect of partnering tools on the contract partnering spirit, and secondly, the effect on the contract elements. Figure 4.2 presents the framework used to evaluate the partnering efforts deployed for TKE Contract 604.

Effect on Partnering Spirit

Successful partnering depends on the endurance of the partnering spirit. It is suggested that among the range of partnering tools employed, their effectiveness in instilling, fostering and maintaining the partnering spirit varies. Instilling a partnering spirit means setting the scene for partnering to foster. Fostering a partnering spirit refers to the strengthening and expansion of the partnering spirit within the programme team. To avoid complacency or reversion back to the traditional confrontational approach, the partnering spirit once established needs to be maintained.

Figure 4.2   An Evaluation Framework for Partnering Efforts

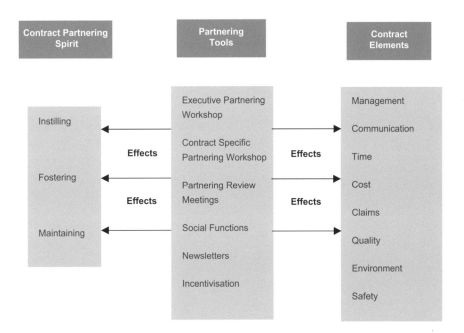

Table 4.1 gives the assessment of the usefulness of the partnering tools in instilling, fostering and maintaining the contract partnering spirit. A total of 16 responses were obtained from MTRC and KG. In addition, eight interviews were conducted to seek views on the effectiveness of the partnering tools adopted in this particular contract. The degree of usefulness is measured on a scale of 1 (no use) to 5 (very useful). The figures provided in Table 4.1 are the averages of the assessment obtained from the interviewees. Discussion on the findings of Table 4.1 is given in the discussion section.

Table 4.1   Usefulness of Partnering Efforts

Partnering Efforts	Usefulness on Contract Partnering Spirit		
	Instilling the Partnering Spirit	Fostering the Partnering Spirit	Maintaining the Partnering Spirit
Executive Partnering Workshop (MTRC staff only)	4.5	3.5	2.5
Contract Specific Partnering Workshop	4.23	3.73	3.8
Monthly Partnering Review Meetings	4.38	4.43	4.68
Social functions	3.85	4.05	3.98
Newsletters	3	4	4
Incentivisation	5	5	5

Effect on Contract Elements

Partnering seeks to bring about a positive impact on contract performance. An interview survey with both MTRC and KG's staff was conducted to seek the degree of usefulness of the overall partnering effort in improving the management, communication, time, cost, claims, quality, environmental and safety issues related to the contract. The degree of usefulness was assessed by a Likert scale of 1 (very negative) to 9 (very positive) with 5 as neutral. Table 4.2 presents the result of the evaluation.

According to the ranking in Table 4.2, the partnering effort was seen as instrumental in reducing claims. Management of the contract ranked second and communication ranked third, followed by cost and time. For these contract elements, the average evaluation scores are all above 7, thus suggesting a positive effect. For contract elements such as quality, environment, and safety, the evaluative score is around 5.5, which is closer to the neutral line. There appears to be a clear distinction between those contract elements positively affected and those where the impact is considered marginal.

Table 4.2   Degree of Usefulness of the Overall Partnering Efforts on
Individual Contract Elements

| Contract Elements | Degree of Usefulness of the Overall Partnering Efforts on Contract Elements | |
	Contract TKE 604	Ranking
Claims	8.2	1
Management	8	2
Communication	7.75	3
Cost	7.65	4
Time	7.31	5
Quality	5.63	6
Safety	5.63	6
Environment	5.34	8

4.9.3   Collection of Empirical Support

As part of the partnering monitoring system, MTRC maintained also records of a number of Key Performance Indicators (KPI) for TKE Contract 604. This stage of the study involved the collection of empirical support for the views expressed by the interviewees. Three main sources were relied on:

  i)   The Monthly Partnering Attribute Scores;
 ii)   The Key Performance Indicators records; and
iii)   The Contract reports.

The monthly partnering attribute score chart (Appendix 1 refers) has already been introduced in the previous section. As for the KPI and Contract reports, MTRC collates and maintains records of quantified measures in relation to the performance of the contract. For example, KPI and contract reports include records in connection with the contract status, such as:

  i)   Programme: sections ahead/behind programme;
 ii)   Cost: variations, commitment to settle;
iii)   Quality: non-conformance reports and time taken to rectify;

iv) Safety: accident types and rates, and severity rate;

v) Environment: prosecutions by the Environmental Protection Department;

vi) Claims: time taken to settle;

vii) Communication: level of correspondence to/from contractor; and

viii) Change: numbers of variations and revised drawings issued.

4.9.4 Effective Partnering Tools in TKE Contract 604

Partnering goes beyond a contract strategy. Other than the procedural arrangements, successful building up of a partnering spirit underpins contract success. The desired partnering spirit can be contract specific and is often expressly stated in the partnering charter. There are three phases in building a partnering spirit, instilling, fostering and maintaining. In this section, based on the interviews and evaluation results (refer to Table 4.1), the three phases in building a partnering spirit in the TKE Contract 604 are discussed.

Instilling Partnering Spirit

The construction industry traditionally has an adversarial culture that is incompatible with the partnering concepts. To overcome this inertia and set the scene for partnering, the commitment and dedication at the top management level must be unequivocal. Initiation has to be top-down and senior management at contract level should be empowered for implementation. To instill a partnering spirit at the contract level, senior management at that level plays a pivotal role. The initial partnering workshop, the holding of monthly review meetings and the collation of control data (partnering attribute scores and KPI's) have sown the seed for the partnering spirit to foster. The TKE Contract 604 was exemplary in this respect, the commitment of key management staff of MTRC and KG was evidenced in their enduring participation in the partnering review meetings as well as the various initiatives to resolve problems. These included the reduction in the level of written correspondence, the encouragement to use informal communication, and the formulation of a problem resolution procedure. To illustrate the improvement on communication, Figure 4.3 shows the number of incoming and outgoing written correspondence over the project duration. Except for a few sharp rises due to the intensity of construction activities prevalent at that time, a downward trend in the level of correspondence (in and out) can be observed. MTRC conducted a study that estimated the unit cost for handling each item of incoming and outgoing correspondence at HK$400 and HK$800 respectively. On this basis the monthly cost of correspondence between meetings 5 and 19 reduced by some HK$400,000.

Figure 4.3   Level of Written Correspondence to and from the Contractor

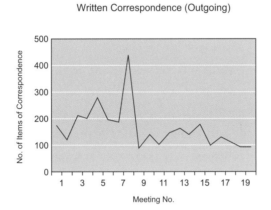

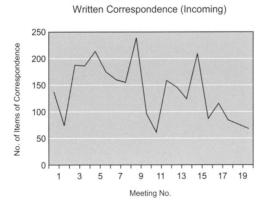

To instill a partnering spirit effectively, senior management must have faith in partnering. In addition, senior management needs to be agile in capitalizing on opportunities upon which the underlying principles of partnering can be manifested, thus enabling the establishment and subsequent achievement of common goals. The results suggest that the MTRC executive partnering workshop sowed the seed for a partnering spirit to grow on the TKE project. However, partnering could only flourish with KG's agreement and commitment. The contract specific workshop and the partnering review meetings were effective tools in instilling the partnering spirit.

Fostering Partnering Spirit

Fostering the partnering spirit includes all steps taken by the management at the contract level to ensure the partnering message infiltrates the organization's hierarchy, together with the actions plans and policies that manifest the partnering spirit. During the interviews, it was highlighted that not only should senior management be well-versed with the partnering principles, it was of equal importance that the working and site level fully understand partnering. As partnering requires a team effort rather than individual input, commitment from all levels were therefore essential. On the TKE Contract 604, review meetings, social functions and newsletters were organized and considered effective in fostering the partnering spirit. With the involvement of the committed and understanding senior management, the team managed to establish good relationships throughout the project duration. In relation to the monitoring of the partnering status, the partnering scores were good measures of the current partnering status. The scores reflect the personal value and judgments of the contract participants. As raised by the senior management from KG, the TKE Contract 604 proved to be a successful partnering venture, despite the many problems encountered. Partnering in construction is often seen as "lip service" without much substance in practice. On Contract 604 both parties agreed that openness and trust was substance to the partnering.

Maintaining Partnering Spirit

To maintain the partnering spirit, there is a need to continually emphasize the partnering ethos and reinforce the partnering principles. It has been suggested that, no matter how good the relationship is, constant monitoring and checking remain important at all stages of contract. This was a view supported by the interviewees. The regular partnering review meeting was designed for this purpose and review workshops were used to reinforce the commitment to the partnering principles and update the common goals and objectives. Relationship difficulties amongst the front line staff were identified and actions were taken to address this. These included junior level workshops, team talks and social events. In general, the partnering spirit of the TKE Contract 604 was maintained at a high level throughout.

Effective Partnering Tools in the TKE Contract 604

According to the feedback obtained from the interviews and the evaluation results, the partnering review meeting and the incentivisation agreement were identified as the two most instrumental partnering mechanisms.

Monthly Partnering Review Meetings

During the review meetings the scores and their changes were discussed; a process through which problems or areas of concern were identified. The meetings served as a valuable platform for grievances to be aired, which otherwise, might have remained buried in the organizational hierarchy and become a source of conflict. Care had to be taken however to guard against the meeting being turned into a dispute adjudication forum.

Partnering review meetings were also regarded as a useful channel for the exchange of important information and ideas. For MTRC, it was a way of learning about KG's view of the contract. For KG, they saw it as an effective means of voicing their concerns and dissatisfaction. In fact, it was through constructive debates that problems could be resolved promptly. Communication was never one-way or one-sided, meetings of this kind helped to maintain active and open communication in the potentially hostile construction contracting environment. The open discussion helped parties to develop trust and establish a communication network. The monthly review meetings together with the partnering score charts proved very useful as a tool to reflect the current partnering status. As an illustration, the average partnering scores of MTRC and KG are presented in Figure 4.4.

Monthly review meetings were held after the contract specific partnering workshop, which sought to instill the partnering spirit. Initially a slight decline in partnering scores was recorded. The partnering spirit had not developed and as construction activities began to grow, problems began to surface. The mutual trust required for partnering to prosper had not yet established. In particular, little change in the partnering scores was recorded by KG. MTRC realized the need to improve the partnering spirit and around the time of review meetings No. 8 and No. 9, negotiation of an incentivisation agreement with KG began. The incentivisation agreement was agreed in principle in December 2000 (meeting No. 10) and signed in February 2001 (meeting No. 12). Since then, a rising trend of average partnering scores was noted.

Incentivisation Agreement (IA)

The IA was regarded as a critical element to the success of the partnering venture in TKE Contract 604. The IA reduced uncertainty and promoted active and open partnering commitment. Both parties strongly believed that the IA was particularly useful in resolving claims. In addition, the interviewees agreed that the success of the IA relied on the shared risk provision, which provided for a target cost set against an agreed list of risks. Any savings against the target cost were to be shared between MTRC and KG. They believe that the IA very useful in bringing the parties together to focus on a common objective. Both MTRC and KG welcomed the use of IA and their satisfaction is reflected by the improved relationship and trust. To illustrate the importance of the IA it is interesting to note the changes in some of partnering scores and the contract elements before and after the conclusion of the Incentivisation Agreement. Figure 4.5 shows the movement of a number of the Partnering Scores: trust, honesty, communication, teamwork and financial objectives of both MTRC and KG. Figure 4.6 presents the record of claims raised by KG, the time to resolve claim and percentage of application actually certified, before and after the IA.

Figure 4.4   Partnering Scores Movement (MTRC & KG)

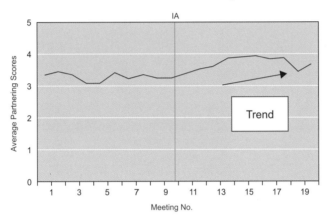

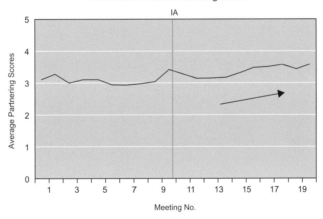

Figure 4.5   Movement Chart of Partnering Attribute Scores (MTRC & KG)

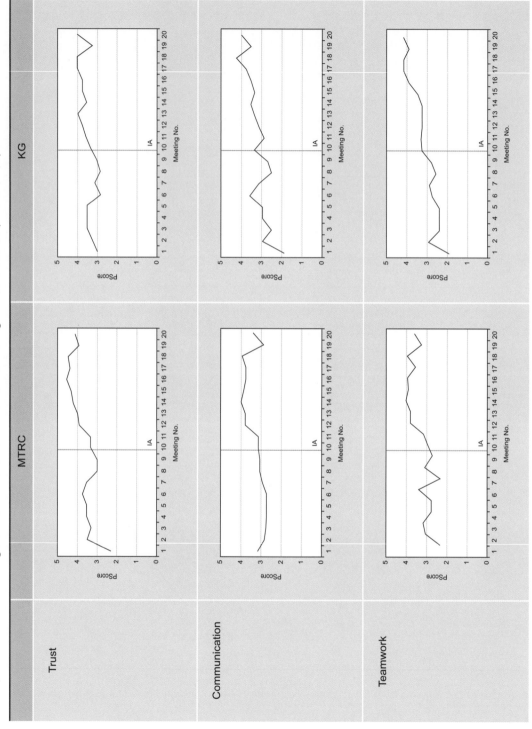

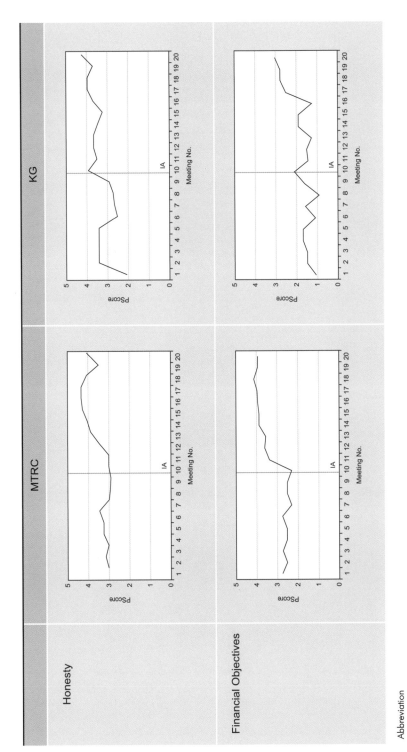

Abbreviation

IA—Time when the terms of Incentivisation Agreement finalized

Figure 4.6   Effects of IA on Claims and Certification

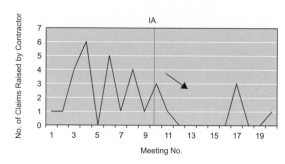

Claims Raised by Contractor

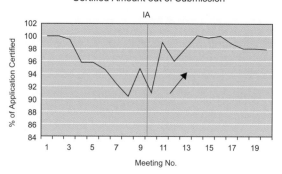

Certified Amount out of Submission

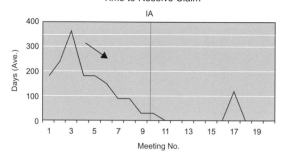

Time to Resolve Claim

Abbreviation:
IA—Time when the Terms of Incentivisation Agreement finalized

KG offered a very competitive tender price and entered into the contract with a negative account projection. Traditionally this could have led to a very aggressive contractual relationship. Indeed, the initial rocky site environment can be evidenced from the low partnering scores from both sides. It was agreed to introduce the concept of incentivisation around the period of review meetings 8 and 9. Incidentally, this coincided with the partnering scores being at an all time low. It is apparent that with the introduction of the IA, the partnering scores began to pick up and showed a steady upward move upon the conclusion of the IA. The positive effect of the IA on the contracting environment is reflected by the rise in scores for trust, honesty, communication, teamwork and financial objectives (refer to Figure 4.5).

The impact of IA on the commercial aspects of the contract is supported by facts that the number of claims submitted reduced significantly as KG was more concerned in not expanding the shared-risk portion of the IA; the percentage of payment submission certified increases; and the time taken to resolve claim reduces.

In addition, the movement of Shared-Risk Cost as illustrated in Figure 4.7 further supports the commercial value of IA in partnering project. These suggest the tension within the contract had been relieved and efforts were being directed to achieve the aligned common goal: minimize expenditure of the shared-risk portion. Cost saving initiatives through value engineering increased the amount of the shared-risk sum. The use of the IA in the TKE Contract 604 was clearly a big step forward towards amicable contracting as the risks were discussed and the allocation agreed between the parties.

4.9.5 Lessons Learnt from MTRC TKE Contract 604

Partnering in construction re-orients the attitude of contract participants from confrontational to co-operative. This requires the commitment of both the contracting parties. To facilitate effective partnering, MTRC TKE Contract 604 employs a range of tools to instill, foster and maintain partnering spirit. The monthly review meetings and the incentivisation agreement are identified as the most effective tools to effectuate the partnering approach. A partnering arrangement advocates a "win-win" approach. The review meetings provided a platform for open communication among the contract participants. The incentivisation agreement for the TKE Contract 604 provided a target for both parties to strive for timely completion and effective working methods and hence a reduction in costs. Commitment lies at the heart of all partnering arrangements and it cannot be sustained if there is no realizable benefit. TKE Contract 604 is a solid example of how partnering can be introduced and maintained.

Figure 4.7   Movement of Shared Risk Cost

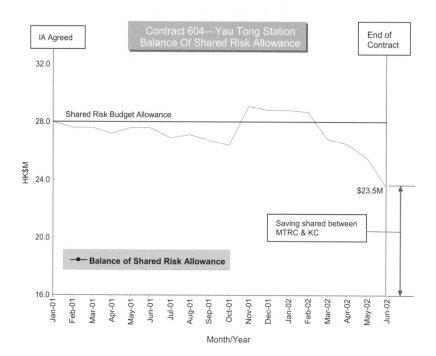

4.10 Summary

Partnering is considered as an exemplar of co-operative contracting in construction. The use of the partnering approach to procure construction project has gained momentum in the last decade. Partnering is now the norm rather than the exception for most projects of reasonable size in both public and private sectors. Partnering arrangement can be project based or strategically oriented. It has been advocated that in order to derive tangible benefits, strategic partnering that builds on established relationship is necessary. Nonetheless, project partnering is inevitable and it is the building block of strategic partnering. This chapter gives an account of the focus of partnering. Experiences accrued in Australia and the U.K. are touched upon then followed by the experience in Hong Kong. A case study on a MIRC project, TKE contract 604, is used to illustrate the essential features and tools used. The case study neatly pointed out the important of a behavioral change in contracting if partnering stands any chance of success.

References

Associated General Contractors of America. 1991. *Partnering: A Concept for Success. Washington.* U.S.A.: Associated General Contractors of America.

Barlow, J., M. Cohen, A. Jashapara and Y. Simpson. 1997. *Towards Positive Partnering.* Bristol: The Policy Press.

Bayliss, R. 2001. MTRC corporation—Progress through partnership. MTRC Ltd. unpublished paper 2001.

Bayliss, R. 2002. Project partnering—A case study on MTRC Corporation Ltd's Tseung Kwan O Extension. *HKIE Transactions* 9 (1).

Bennett, J., and S. Jayes. 1995. *Trusting the Team: The Best Practice Guide to Partnering in Construction.* Reading: Centre for Strategic Studies in Construction.

Bennett, J. and S. Jayes. 1998. *The Seven Pillars of Partnering.* London, UK: Thomas Publishing Ltd.

Black, C. 1999. Operational risks associated with partnering for construction. In *Profitable Partnering in Construction Procurement,* ed. S. O. Ogunlana. London, UK: E & FN Spon.

Bresnen, M., and N. Marshall. 2000. Partnering in construction: A critical review of issues, problems and dilemmas. *Construction Management and Economics* 18(2): 229–237.

Butcher, T. 1997. Partnering: Contractual considerations. *Construction Law* 9(3): 79.

Cheung, S. O., S. T. Ng, S. P. Wong, and C. H. Suen. 2003. Behavioral aspects in construction partnering. *The International Journal of Project Management* 21(5).

Construction Industry Institute. 1989. *Partnering: Meeting the Challenges of the Future.* Texas, U.S.A.: Construction Industry Institute.

Construction Industry Institute Australia. 1996. *Partnering: Models for Success.* Construction Industry Institute.

Construction Industry Institute. 1991. In search of partnering excellence. *Special Publication* 17(1).

Construction Industry Review Committee. 2001. *Construct for Excellence: Report of the Construction Industry Review Committee.* Government of the Hong Kong Special Administrative Region.

Cook, E. L., and D. E. Hancher. 1991. Partnering: Contracting for the future. *Journal of Management and Engineering* 6 (4): 431–446.

Crowley, L.G., and M. A. Karim. 1995. Conceptual model of partnering. *ASCE Journal of Management in Engineering* 11: 33–39.

Drexler, J., and E. Larson. 2000. Partnering: Why project owner-Contractor relationship change. *Journal of Construction Engineering and Management* (July): 293–297.

Egan, Sir J. 1998. *The Egan Report—Rethinking Construction.* UK: HMSO.

Glesson. 2005. Glesson is at the forefront of partnering initiatives in the UK construction industry. http://www.mjgleeson.com

Gransberg, D., W. Dillon, L. Reynolds, and J. Boyd. 1999. Quantitative analysis of partnered project performance. *Journal of Construction Engineering and Management* 125 (3): 161–166.

Gyles, Q. C. 1992. *Royal Commission into Productivity in the Building Industry in New South Wales.* The Government of the State of New South Wales, Australia.

Handy, C. 1993. *Understanding Organizations.* 4th ed. Harmondsworth: Penguin.

Hartnett. 1990. Partnering. *The Military Engineer.* U.S.A.: The United States Army Corps of Engineers.

Hellard, R. B. 1995. *Project Partnering: Principle and Practice.* London, UK: Thomas Telford Publications.

Holti, R., and H. Standing (1996). *Partnering as Inter-related Technical and Organizational Change.* London: The Tavistock Institute.

Hong Kong Housing Authority. 2000. *Quality Housing, Partnering for Change*. The Government of the Hong Kong Special Administrative Region.

Kwan, A., and G. Ofori. 2001. Chinese culture and successful implementation of partnering in Singapore's construction industry. *Construction Management and Economics* 19(6): 619–632.

Larson, E. 1997. Partnering on construction projects: A study of the relationship between partnering activities and project successes. *IEEE Transactions on Engineering Management* 44(2): 188–195.

Latham, M. 1994. *Constructing the Team—Final Report of the Government/Industry Review of Procurement and Contractural Arrangements in the UK Construction Industry*. UK: HMSO Department of the Environment.

Li, H., E. Cheng, and P. Love. 2000. Partnering research in construction. *Engineering, Construction and Architectural Management* 7(1): 76–92.

National Economic Development Council's Construction Industry Sector Group (NEDO). 1991. *Partnering: Construction without Conflict*. NEDO. London.

Masters Builders Australia (MBA). 2005. *The Effectiveness of Partnering*, ed. Denis Wilson (National Director, Training and Partnering). http://www.leadr.com.au/WILSON.PDF

Newman, P. 2000. Partnering, with particular reference to construction. *Arbitration Journal* 66 (1): 39.

Office of Government Commerce (OGC). 2003. *Strategic Partnering Taskforce*. http://www.odpm.gov.uk/pub/521

Peace, S., and J. Bennett. 2002. *How to Use a Partnering Approach for a Construction Project: A Client Guide*. The Chartered Institute of Building: Englemere.

Smyth, H. 2002. Partnering: Practical problems and conceptual limits to relationship marketing. *International Journal for Construction Marketing*. http://www.brookes.ac.uk/other/conmark/IJCM/issue_02/010202.html

Thompson, P., and S. Sanders. 1998. Partnering continuum. *Journal of Management in Engineering* 14(5): 73–8.

Appendix 1— Partnering Scores Chart

Partnering Monitoring Chart —March XXXX

Objectives/ Values/ CSFs	A	B	C	D	E	F	G	H	I	J	K	L	Average	Last Month
1. Trust														
2. Honesty/Openness/Integrity														
3. Communication														
4. Relationship/Teamwork Co-operation														
5. Programme														
6. Quality														
7. Safety														
8. Financial Objectives														
9. Positive/Respect/Job Satisfaction														
10. Competent and Sufficient Resources/Commitment														
11. Waste Minimization														
12. 3rd Parties Needs														
13. Problem Resolution Process														

1 = Strongly agree
2 = Slightly agree
3 = No feeling either way
4 = Slightly agree
5 = Strongly agree

Trust: Foundation of Co-operative Contracting

Sai On Cheung
Wei Kei Wong
Peter Shek Pui Wong
Henry Suen

Acknowledgements
Part of the content of this chapter has been published in Vol. 21 of the *International Journal of Project Management*. The authors thank the journal for the permission to reproduce the content therein in this Chapter.

5.1 Introduction

Relational Contracting theory can be viewed as a pragmatic view on contracting practices. In Chapter three, the transaction characteristics of relational type of contract have been discussed. These include co-operation, culture, risk, good faith, flexibility, dispute resolutions and duration. The reservation over the existence of a relational contract law mainly stems from the lack of backing by established legal principles and doctrines. Nonetheless, it is suggested that the most critical factor to manifest co-operation in contracting is trust and the chapter is devoted to deliberate this proposition.

5.2 Trust: The Foundation of Co-operative Contracting

Unbalanced risk allocation in contracts, adversarial relationships between project participants together with the traditional client-contractor mentality, these have long been identified as the causes of problems in the construction industry (Jannadia *et al.* 2000). Practitioners in the field share the view that contract provisions tend to favor the clients, leaving all the burdens on contractors and are often rigidly interpreted without taking into account the parties' needs and practical difficulties (Piper 2001). The problem is further complicated by traditional contracting systems in which the liabilities of the client and contractor are explicitly stated as terms of a contract. In situations where the extent of risk is uncertain, the contractor is usually the one dumped with the risk, simply for administrative convenience. This arrangement hinders dispute settlement as parties tend to revert to their contractual positions when problems occur. As a consequence, open and honest communications are rare while confrontation and adversarial relationships are common in the industry.

The breakdown of relationships among project team members is extremely common in the construction industry. As project team members typically belong to different organizations, skepticism and conflict of interest seem a natural course. This is exacerbated by the increasing competitive construction market. If these doubts are allowed to grow and fester, communication between project team members will be blocked and creativity will be stifled (Beccera and Huemer 2000).

Communication will shut down because when project team members feel they have been taken advantage of. They would then stop dialogue and refrain from meeting one another. This can have a very negative effect on the project, because when project team members stop communicating, ideas stop flowing as well (Mayer *et al.* 1995). Creativity will also be stifled because when project team members do not trust each other, they become risks aversion and tend to maintain status quo. Without risk-taking, creativity can hardly be fostered. This would cripple the functioning of the project team if members are afraid to make suggestions for project improvements.

Like a marriage, it takes two parties to manifest "trust". In their studies, Wood and McDermott (1999) defined trust as "a willingness to rely on the actions of others, to be dependent upon them, and thus be vulnerable to their actions". It is now widely accepted that trust is the foundation and one of the critical success factors for project partnering (Law 2004, McDermott *et al.* 2005). Nooteboom (1999) pointed out that "Trust-based partnering is characterized by honest/open communications, promise keeping, an absence of deception, fair sharing of benefits and respect and reciprocity. All this is best seen in the need for relationships characterized by reciprocity". As discussed in Chapter four, partnering is an exemplar of co-operative contracting in construction. Hence, in this chapter, the terms partnering and co-operative contracting are used interchangeably.

To establish trust in a partnering project, several issues have been identified as essential (Brenkert 1998):

(1) Project team members must acknowledge that trust is important. They must feel comfortable in discussing solutions for problems and share their experiences.

(2) Feedback system on trust must be provided, i.e., keeping record of how members think of their "level of trust" on a regular basis, which then followed by ways to build trust in their workplace.

(3) Make trust a measurable component of project success. It is important to gauge whether trust is actually building within the project team. Senior management should maintain an open attitude in addressing team members' concerns and consider their suggestions.

The partnering efforts of TKE Contract 604 have been described in Chapter four. Employing the same case study, the significance of the behavior aspects of partnering are highlighted in this chapter.

5.3 The Experience of TKE Contract 604

To make partnering effective, it is critical to have a change in culture within the industry. This can only be achieved with the change in attitude of the project participants. This section discusses the importance of this behavioral aspect of partnering by analyzing the data collected from the TKE Contract 604. Partnering is not easy to define. In the area of supply chain management, partnering is framed with a relationship paradigm. Table 5.1 presents a summary of partnering critical success factors (Kumaraswamy and Matthews 2000, Li *et al.* 2000). It can be observed from Table 5.1 that many of the success factors are behavioral or attitudinal in nature. These include trust, co-operation, concern for relationship, and commitment. Partnering is just lip service if the attitude of the participants remains adversarial and co-operation cannot be expected. Project performance is often scarified because of the many behavioral traits that prohibit co-operation (Bayliss 2002). Such a view is also shared by the first ever review on the Hong Kong construction industry: "the industry is very fragmented and is beset with an adverse culture. Many industry participants adopt a short-term view on business development, with little interest in enhancing their long-term competitiveness" (CIRC 2001). Partnering manifests co-operative contracting where information and risks are shared as appropriate. This revolutionary approach has received heavy skepticism, as it appears that partnering seeks to align objectives of the contracting parties, which are inherently incompatible, if not opposite.

Table 5.1   Partnering Success Factors

Li, Cheng and Love (2000)	Devilbiss and Leonard (2000)	Construction Industry Institute Australia (1996)	Matthews (1996)
• Adequate Resources • Management Support • Mutual Trust • Long-term Commitment • Coordination • Creativity • Effective Communication • Conflict Resolution	• Trust • Understanding of Partner's Needs • Conflict Resolution • Faith	• Commitment • Trust • Equity • Mutual Goals and Objectives • Implementation • Joint Process Evaluation • Dispute Resolution Process	• Goals and Objectives • Trust • Problem Resolution • Commitment • Continuous Evaluation • Group Working • Win-Win Philosophy • Shared Risk • Equity • Co-operation
Reading Construction Forum (1995)	**Sanders And Moore (1992)**	**Associated General Contractors America (1991)**	
• Free and Open Communication • Open Book Costing • Annual Review of Performance • Workshops • Continuous Evaluation • Mutual Objectives • Problem Resolution	• Co-operative Management Team • Co-operation • Open Communication • Group Working • Common Goals • Problem Solving	• Commitment • Continuous Evaluation • Equity • Mutual Objectives • Timely Responsiveness • Trust • Implementation	

Source: modified from Kumaraswamy & Matthews (2000) and Li *et al.* (2000)

Is there an overriding attitudinal factor pivotal to a partnering venture? Moore (1999) suggested that partnering is about management of relationship that must be trust-based. Nonetheless, Larson and Dexler (1997) critically pointed out that the partnering literature is relatively quiet on how trust element can be developed in partnering.

5.3.1 Trust: The Pivotal Attitudinal Factor

As mentioned in the above, the type of co-operation envisaged in a partnering setting may not be readily available due to the inherent conflicting objectives of the contracting parties. However, this is only true if no benefit can be derived from being co-operative. If benefits exist, the key question becomes how can these be made known to the parties? This type of problems falls into the classic work of Rapoport *et al.* (1965) described as prisoner's dilemma, which refers to a two party non-constant–sum game in which some outcomes are preferred by both parties to other outcomes. The moves of the parties can either be competitive or co-operative. Competitive moves are those that focus on one's own interest. This generally would invoke retaliation or defensive responses. Co-operative moves are those that put the interest of both parties first. Co-operative moves are characterized by reciprocal moves. That means if A is behaving co-operatively, he is expecting a co-operative response. This expectancy lies in the central heart of trust. Trust is an attitude of human acts or beliefs. It is a complex construct with multiple bases, levels and determinants (Rousseau *et. al* 1998). Furthermore, trust need to be earned. Trust can only be built on a strong degree of predictability. That means one party can comfortably assume that for a co-operative move he made, a reciprocating co-operative move will be returned.

This expectancy and predictability dimensions of trust are well supported by other studies. According to Nyhan and Marlowe (1997), trust is often associated with situations involving personal conflict, outcome uncertainty, and problem solving. It is an expectation and a prediction of future events. Varying in intensity, trust is associated with the confidence in and reliance upon the prediction (Bernstein *et al.* 1989). Trust represents a favorable interpersonal or inter-organizational relationship and is an important component in the long-term stability of the members of the organization (Cook and Wall 1980, Hart 1988). Trust is dynamic, always either growing or diminishing. The happenings during the course of a project would intensify or diminish the level of trust among the people involved (Hawke 1994). Rotter (1967) proposed that interpersonal trust is an expectancy held by an individual or a group that word, promise, verbal or written statement of another individual or group can be relied upon. According to his definition, trust is not related to specific experience but generalized expectancy derived from experiences that individuals perceive (Jones and George 1998). Ford (2002) provides a summary of trust definitions characterized by predictability of human behavior (refer to Table 5.2).

Table 5.2   Trust Defined by Predictability on Human Behavior

Definition of Trust	Citation
"Trust is a psychological state comprising the intention to accept vulnerability based upon positive expectations of the intentions or behavior of another."	Rousseau *et al.* (1998)
"Trust is a psychological construct, the experience of which is the outcome of the interaction of people's values, attitudes, and moods and emotions."	Jones & George (1998)
Trust is the "expectation of regular, honest, and co-operative behavior based on commonly shared norms and values."	Doney, Cannon & Mullen (1998)
"Trust is the degree to which the trustor holds a positive attitude towards the trustee's goodwill and reliability in a risky exchange situation."	Das & Tang (1998)
"Trust exists in an uncertain and risky environment; trust reflects an aspect of predictability—that is, it is an expectance."	Bhattacharya, Devinney & Pillutla (1998)
Trust is "one's expectations, assumptions, or beliefs about the likelihood that another's future actions will be beneficial, favorable, or at least not detrimental to one's interests."	Robinson (1996)
Trust is "the willingness of a party to be vulnerable to the actions of another party based on the expectation that the other will perform a particular action important to the trustor, irrespective of the ability to monitor or control that other party."	Mayer, Davis & Schoorman (1995)

Source: modified from Ford 2002

5.3.2 The Roles of Mistrust

Another way to look at trust is from the perspective of mistrust. According to Whitney (1999), the five main sources of mistrust include:

Misalignment of Measurements and Rewards

This refers to the situation that mistrust is generated as a result of the mismatch of effort and reward. This may arise firstly if the measurements of effort are set of the mismatch of effort and reward. This may arise firstly if the measurements of effort are set on something that is either peripheral or dysfunctional to the objective. Secondly, even with a reasonable measurement as well as the reward for the effort, if not aligned, mistrust will result. Measurement of effort is rare in construction. Instead, parties to a construction contract often take a very inflexible approach in their contractual positions, in particular with regard to their rights. Hence, parties typically take a very strict view on responsibilities and under value the interdependent nature of construction activities. This works against the appreciation of the other's effort. This ingrained attitude negates the call for co-operation and explains why incentive is so rare in construction contracting parties.

Incompetence

According to Whitney (1999), this is the most costly of all causes of mistrust. For example, when we hire people who lack the necessary knowledge and skills, instead of helping them correct their deficiencies, we supervise and inspect them. The cost is enormous considering the layers of supervision and the concomitant suboptimisation that springs from a dispirited work force. In other instances we just assume that people are incompetent without giving them a real chance to show what they can do. The potential negative impact on the workforce is these people may direct their energy elsewhere and in extreme cases create conflict and trouble. The analogy in construction is that the employer and consultants tend to think that whatever problems cropped up on site are the result of the incompetence of the contractor. Tedious inspection systems are commonly installed as control measures. The cost involved is enormous. This misconception is a source of mistrust.

Lack of Appreciation of a System

One needs to realize that a chain is as strong as its weakest link. Hence no part of a system should consider they are more important than the other when the success of a system hinges on the mutual dependence of the sub-systems. The lack of appreciation of this interdependency would severely limit the development of trust within the system. Construction projects are good examples of such system. Mistrust germinates where mutual dependence is not respected.

Untrustworthy Information

When information that project participants are expected to act on is incomplete, biased or wrong, trust is always at risk because defensive responses are triggered. It is not uncommon in construction that information is withheld either totally or in part, typically with two intents: to hide deficiency or to deceive the other. When these happen, trust development will be severely hampered.

Failure of Integrity

Failure of integrity involves lying, cheating and/or stealing. This may happen on individuals or the organizations as a whole. Where integrity fails, trust will be tarnished. Partnering in construction advocates co-operation which cannot be achieved if the contracting parties do not trust each other. From this perspective, mistrust lies in the heart of the potential blockages against partnering. Construction industry involves a large number of participants with different interests thus flourishing a "blame" and "uncompromising" culture. These all work against the partnering spirit.

5.3.3 Developing Trust in Construction Partnering

It is the conscious thinking process which people use to evaluate object. That is, the choice as to whom we trust depend on whether there are good reasons constituting evidence of trustworthiness (Lewis and Weigert 1985). People who make emotional decisions based on trust relationships, express genuine care and concern for the welfare of partners, believe in the intrinsic virtue of such relationships, and anticipate that these sentiments are reciprocated (Black *et al.* 1999, Lewis and Weigert 1985, McAllister 1995, Pennings and Woiceshyn 1987). Frequent interactions are positively associated with the development of trust (McAllister 1995). When project participants continually meet one's trust expectations, it is only natural to develop an affinity for the other as a truth-teller. In other words, when the individual has a series of expectations met, trust will then be developed (Ford 2002). The trust so developed is an emotional bond between the contracting parties. This type of trust can further be nourished and strengthened through repeated co-operative interactions (McAllister 1995).

To decide whether the contractor is trustworthy, developers typically use assessment criteria like: meeting project targets, reasonableness of claims and comments from consultants. Likewise, prompt responses on payment application, attitudes on claims negotiations affect the trustworthiness of the employer in the eyes of the contractor. Hence during the construction stage, if the contracting parties' words and behaviors are reliable, consistent and helpful, even faced with an uncertain or risky situation, the trust among the contracting parties will be developed (Matthai 1989). In general, a construction project will last for several years. It is expected that with positive outcomes from continued dealings between developers and contractors, their relationships will be reinforced by an emotional bond (Lewis and Weigert 1985). Effective communication and interaction among parties are essential for parties to understand others' needs and difficulties (Lewis and Weigert 1985).

To accomplish the partnering endeavors, the MTRC in its TKE Contract 604 had developed a comprehensive partnering implementation programme. The details of the tools used can be found in Chapter four. For the purpose of this chapter, the mechanisms of the monthly partnering review meeting are elaborated. In addition to the initial effort to put partnering in perspective, monthly review meetings are arranged to monitor the partnering status of the project. For this purpose, MTRC designed a partnering score questionnaire as shown in Figure 5.1. Principally, thirteen partnering attributes, addressing both the soft and hard issues of the project, are monitored. The soft issues are mostly those related to behavioral aspects such as trust, honesty, communication, relationship and job satisfaction. The hard issues address the project specific issues such as programme, quality, safety, financial, resources, waste, other parties' needs and problem resolution. Prior to each meeting, participants were required to assign scores (on a scale of 1-strongly disagree to 5-strongly agree), representing their views on the degree of achievement of the thirteen partnering attributes as stated. Being honest and self-critical is paramount in the score assignment process. During the meetings, the scores and their changes are discussed, a process through which problems and areas of concern are unveiled. The changes in the scores can be used as a reflection of the achievement of a certain partnering attribute in the preceding month. A decline suggests something negative might have been happened and the awareness of management team is thereby raised. The meetings serve as a valuable platform for grievances to be aired, which otherwise, might then be buried in the organizational

Figure 5.1   Questionnaire for the Partnering Attributes Scores

Company : MTRC

Contract: 604

Name (Optional):

Month: March   Year: 2000

Please complete this questionnaire and return it to _____ by _____. Complete those sections relevant to you and consider only your involvement with this single contract, and your relations with other people only in so far as they are connected with this contract.

The questionnaire consists of a series of statements to which you are asked to indicate how much you agree or disagree.   Please tick the appropriate box to indicate how you feel at present.

	Strongly Disagree (1)	Slightly Disagree (2)	No Feeling Either Way (3)	Slightly Agree (4)	Strongly Agree (5)
1. I feel I am working in an environment of trust	☐	☐	☐	☐	☐
2. I feel that in the working relationships between all individuals are honesty, openness, and integrity.	☐	☐	☐	☐	☐
3. I feel that good communications are being maintained	☐	☐	☐	☐	☐
4. I feel that I am working in a united team where relationships are good	☐	☐	☐	☐	☐
5. I feel that the contract is achieving the objectives of the programme.	☐	☐	☐	☐	☐
6. I feel that the works are being carried out to the expected quality.	☐	☐	☐	☐	☐
7. I feel that the team actively considers safety and that the works are being carried out safely.	☐	☐	☐	☐	☐
8. I feel that my organization is achieving reasonable commercial success from this contract.	☐	☐	☐	☐	☐
9. I feel I am working in a positive and enjoyable atmosphere, being respected and having job satisfaction.	☐	☐	☐	☐	☐
10. I feel that all parties are deploying competent and sufficient resources for the Contract.	☐	☐	☐	☐	☐
11. I feel that all parties are attributing to minimizing all forms of waste from design, construction and interface perspectives.	☐	☐	☐	☐	☐
12. I feel that the influence, requirements and needs of 3rd parties are being properly considered.	☐	☐	☐	☐	☐
13. I feel that disputes are being resolved in accordance with the agreed problem resolution mechanism.	☐	☐	☐	☐	☐

hierarchy and subsequently generate conflict. It is however necessary to avoid the meeting becoming a dispute adjudication forum

MTRC also installed other monitoring devices, key performance indicators and contract reports in addition to the partnering scores questionnaire. The data/information was collected on a regular basis, usually monthly, for use by management to monitor the TKE project. The data/information is not only useful for project management. In Table 5.1, the success factors for partnering, as identified in reported studies, are presented. It has been suggested that many success factors are related to the attitude of the participants.

5.3.4 Measuring the Partnering Status

The partnering attribute score questionnaire (refer to Figure 5.1) is a useful tool to record the partnering status. The thirteen partnering attributes used in the questionnaire can be categorized into behavioral and project specific. The behavioral attributes are described as attitude oriented. For the project specific group, the attributes are further subdivided into performance oriented or process oriented. Table 5.3 presents the grouping of the partnering attribute scores.

Table 5.3   Grouping of Partnering Attribute Scores

Soft (Behavioral) Issues	Hard (Project Specific) Issues	
Attitude-Oriented	Performance-Oriented	Process-Oriented
• Trust	• Programme	• Resource Commitment
• Honesty/Openness/Integrity	• Quality	• Waste Minimisation
• Communication	• Safety	• 3rd Parties Needs
• Relationship/Co-operation	• Financial Objective	• Problem Resolution Process
• Job Satisfaction		

TKE Contract 604 commenced in late 1999 and up to December 2001, the time when this study was conducted, twenty sets of partnering attributes scores were recorded. In addition to the partnering attribute scores, MTRC also kept track of the contract performance by compiling Key performance Indicators (KPI) and Contract Reports. These records provide valuable performance measures as the contract progresses. Indicators reflecting the communication, time, cost, claims and quality aspects of a contract were short-listed for this study, firstly because of their availability and secondly because of the numeric nature of the data. These indicators are described as contract elements, detailed descriptions and measurements are given in Table 5.4.

Table 5.4   Contract Elements and Their Measurements

Communication
• Average turnaround time for submissions in the period _____days
• Average no. of letters / ERFs issued to contractor /consultant _____No.
• No. of letters/ CSFs received from contractor/consultant _____No.
• No. of drawings issued to contractor _____No.
• No. of drawings revised _____No.
• No. of drawings amendment issued to contractor _____No.
• No. of meetings held with contractor/consultant in the month _____No.
• No. of RFIs raised by Contractor _____No.
Construction time
• % completion on time _____ %
Cost
• Number of variations issued: Accumulative total _____No.
• Positive / negative change to contract sum arised from variations _____% + or -
• Certified variation amount out of submission amount _____%
Claims
• Number of EOT claim notices _____No.
• Number of EOT claim submitted _____ No.
• Number of EOT claim resolved _____No.
• No. of cost claim notices _____No.
• No. of cost claim submitted _____No.
• No. of cost claim resolved _____No.
• Dollar value of claims out of contract sum _____%
• Certified claim amount out of submission amount _____%

Table 5.5 R² of Regression Equations between Contract Elements and Partnering Attribute Score Groups

Contract Elements		MTRC Partnering Score Group		
		Attitude (At)	Performance (Pt)	Process (Pr)
Communication	1. Turnaround Time for Submission	0.413	0.342	0.217
	2. No. of EFI Issued	0.489	0.423	0.366
	3. No. of Drawings Issued to Contractor	0.471	0.244	0.371
	4. No. of Drawings Revised	0.481	0.291	0.389
	5. No. of DAM Issued	0.355	0.532	0.192
	6. No. of RFI Raised by the Contactor	0.508	0.464	0.569
	7. No. of Meetings Held with the Contractor/Consultants	0.468	0.318	0.386
	8. No. of Letters/ERF Issued to the Contractor/Consultants	0.382	0.309	0.358
	9. No. of Letters/CSF Received from the Contractor/Consultants	0.473	0.305	0.348
Time	10. % Achieved on Time	0.116	0.169	0.231
Cost	11. No. of Cost Variation	0.305	0.227	0.115
	12. Change of Contract Sum Arising from Variation	0.286	0.113	0.24
	13. Certified Variation Amount Out of Submission	0.413	0.542	0.426
Claims	14. No. of EOT Claim Notices	0.239	0.043	0.065
	15. No. of EOT Claim Submitted	0.363	0.257	0.375
	16. No. of EOT Claim Resolved	0.26	0.62	0.088
	17. No. of Cost Claim Notices	0.239	0.042	0.065
	18. No. of Cost Claim Submitted	0.4	0.422	0.137
	19. No. of Cost Claim Resolved	0.223	0.572	0.131
	20. Dollar Value of Claims Out of Contract Sum	0.213	0.13	0.159
	21. Certified Claim Amount Out of Submission Amount	0.344	0.498	0.357

Regression analyses were performed with each contract elements (as dependent variables) and each category of partnering attribute scores (as independent variables). Table 5.5 presents the summary of the R² of the regression equations. It is to be noted that other than the value of R^2, the prediction ability of a partnering attributes has to be examined by both the sign and magnitude of its regression equation coefficient. In this study, the emphasis is on the identification of the group of significant partnering attributes, the regression equation coefficients of the partnering attributes are not discussed.

Table 5.6   Significant Pearson Co-relation Coefficients between Partnering Scores (Job Satisfaction and Others)

Partnering Attribute Score	MTRC Pearson correlation coefficient Job Satisfaction	Average
Trust	0.812**	Attitude 0.818
Honesty	0.849**	
Communication	0.755**	
Relationship	0.857**	
Programme	0.060	Performance 0.093
Quality	-0.229	
Safety	-0.253	
Financial Objectives	0.795**	
Resource Commitment	0.577**	Process 0.581
Waste Minimization	0.582**	
3rd Parties' Needs	0.504*	
Problem Resolution Process	0.662**	

** Correlation is significant at the 0.01 level (2-tailed)
 * Correlation is significant at the 0.05 level (2-tailed)

It can be observed from Table 5.5 that for contract elements reflecting communication, the R^2 of the regression equations derived from the attitude oriented partnering attribute scores are generally higher than the other two groups (seven out of nine). For contract elements on time, process oriented partnering scores gives the highest R^2. As for cost and claims, the attitude group again displays higher R^2 generally (six out of eleven). It is interesting to note also that these six contract elements are related to submission and/or notices which understandably are more affected by attitude. In addition to the Regression analyzes, the co-relations among the partnering attribute scores are also examined. Table 5.6 presents the Pearson co-relation coefficients between the partnering scores of job satisfaction and the other twelve attributes. This serves to identify the group of partnering attributes scores with high correlation with job satisfaction. It is apparent from Table 5.6 that the attitude oriented group of partnering attribute scores gives the highest co-relation with the job satisfaction score among the three groups of partnering attribute scores.

5.3.5 Key Observations from the TKE Contract 604

Construction activities require concerted effort of all participants. Teamwork is the key. This teamwork can either be intra-organizational or inter-organizational or both. Organizational interest, professionalism as well as individual behavior, singularly or collectively work against team building. Partnering advocates co-operation, open communication and joint problem solving. Its success depends on the attitude of those involved. Partnering has no prospect without the commitment and trust among the participants. The game of prisoner's dilemma highlights the criticality of co-operative moves in the development of mutual trust in non-constant sum game such as construction projects (Rapoport *et al.* 1965). The common goal of the participants includes successful completion of the projects. To bring out genuine co-operation, proactive moves of the parties are necessary in order to break through the inherent mistrust. Revisiting the five sources of mistrust described by Whitney (1999), the following discusses how partnering tools used in the TKE Contract 604 make an impact on trust development. Figure 5.2 provides the framework.

Figure 5.2 presents the five sources of mistrust, the partnering tools and the purposes of these tools in relation to trust building. The contract specific partnering workshop is instrumental in establishing aims, mission, vision and values of the contract. This is further crystallised in writing through the formulation of the contract partnering charter. This sets the scene for trust building. The partnering review meetings have proved to be indispensable in TKE Contract 604. Through these meetings, a contingent organization structure and interactive management process is created. Partnering review meetings establish a channel for regular exchange of important information and ideas. For MTRC, it is a way of learning about the contractor's view of the project. For the contractor, it is an effective means to voice out its concerns. In fact, it is through constructive dialogue that problems are resolved. Communication is never one-way or one-sided, meetings of this kind help to maintain active and open communication in the potentially hostile construction contracting environment. The open discussion helps to develop trust and establishes a communication network. In TKE Contract 604, the chairmanship of the meeting is not "fixed", but rotated amongst suitable participants. This provokes participation and avoids any perception of dominance. In addition, informal gathering such as social functions widen the communication network.

Figure 5.2   Developing Trust in Construction Partnering

Sources of Mistrust	Partnering Tools	Purposes	
• Misalignment of measurements and rewards • Incompetence of the presumption of incompetence of peers or counter-part. • Imperfect understanding of systems, causing activity that diverts effort from the organization's goals. • Information that is biased, late, useless, or wrong. • Lack of integrity	• Project Specific Partnering Workshop • Partnering Charter • Partnering Review Meetings • Social Functions • Partnering Newsletter	• Establish the aim, mission, vision, and values of the project, then manage as an inclusive system. • Create a permeable organization structure and interactive management process in an open, trusting environment. • Understand and communicate the interdependence of all components. • Conduct an audit of formal and informal measurements and controls. • Reinforce the partnering message through report of success stories.	**Trust**

Trust develops with reciprocating co-operative moves. This often arises at times of crisis or problems. Where crisis or problems get resolved with the effort of the other party, trust between the parties grows. This further brings out the message of inter-dependence that is often not respected in construction. Complacency stifles trust building; hence monitoring the partnering status is a must. Those contract specific measures can be achieved through the compilation of key performance indicators and contract reports. As for the behavioral issues such as trust and communication, the partnering attribute scores questionnaire is instrumental. The publication of a partnering newsletter also helps to reinforce the co-operative emotional bond.

Traditional confrontational style of construction management is becoming out of place. Co-operative teamwork offers greater opportunity to achieve project objectives. This requires the project participants understanding the need to initiate this paradigm shift. The experience of the TKE Contract 604 shows that trust is the pivotal attitudinal factor and that trust building is an indispensable exercise of any partnering arrangements. Successful partnering relies on parties' mutual understanding, honesty, openness, and good communication. Such a view is nicely demonstrated by the prisoner's dilemma concept. With the aid of the data collected from the MTRC TKE Contract 604, the importance of the behavioral aspect of construction partnering is empirically supported. The partnering tools used in that contract and their effects on trust-building are described. This suggests that the review meetings and the partnering attribute score questionnaire are extremely useful in developing trust among the project participants but it must be founded on an open and honest communication between the project participants in a partnering setting.

5.4 Trust and Its Delineation in Construction

Confrontation between client and contractor has been identified as one of the major attitudinal problems of the construction industry for quite some times. This undesirable contracting environment has demonstrated to have negative impact on project performance. Notwithstanding, completion of development projects require the collaboration of contracting parties. These include the client team consists of client, architects, engineers, and quantity surveyors as well as the contracting team comprising main contractors, subcontractors and suppliers. Their efforts, pulled together, enable the accomplishment of complex construction projects. These construction projects required the coordinated efforts of the teams, thus involving intra- and inter-organizational co-operation. The adversaries and uncertainties inherent with construction projects have driven the unhealthy growth in tension among team members and often lead to unsatisfactory performance. The undesirable effects caused by the adversarial relationship prompted the need to investigate possible methods to improve the relationship between client and contractor. It is believed that the existence of trust may alleviate this unfavorable relationship. A number of studies have suggested the needs and benefits of trust in some specially arranged construction projects such as partnering based projects (Wong *et al.* 2000; Bayliss *et al.* 2004). Nonetheless, how trust lubricates inter-organizational relationship in construction has yet been investigated. This paper introduces the characteristics of inter-organizational relationship and the contribution of trust in such relationship. As construction projects are delivered by a multitude of organizations, interactions among them are frequent, thus the importance of trust in fostering such co-operation is apparent.

5.4.1 Organizational Relationship

There are two forms of organizational relationship; intra and inter. The distinction between intra- and inter-organizational relationships is whether an individual's relationship is built under the same or different organizations. Although the definition of relationship simply refers to a state of affairs between people, it has been advocated that the operational efficiency of an organization can be reflected by the behavior of its staff (McAllister 1995, Bhattacharya *et al.* 1998, Whitener *et al.* 1998, Das and Teng 2004). This can be explained by the fact that organizational behavior is manifested by the collective behavior of the individuals involved. In this context, the importance of relationship becomes significant. Organizational behavior contains 3 "how"s: 1) how people perform in their organizations; 2) how people treat others in either the same company or different corporations; and 3) how people respond to others' behaviors. It is a systematic study and application of knowledge about how people, individuals, and groups demonstrate themselves in organizations. A person's behavioral style changes according to variation of the environment surrounding him. Figure 5.3 presents how relationship is manipulated within an organization. In an organization, its culture can be expressed by organizational systems and organizational goals, while members of an organization are brought together through communication. Organizational culture moulds the behavior of its members. As far as culture is concerned, organizational systems are characterized by procedural fairness. In addition, organizational culture is also dependent on the degree of goal attachment of its members. Therefore, individuals' behavior is dynamically affected by the organizational culture, communication, procedural fairness, and adherence to organizational goals (Carnevale and Wechsler 1992).

Figure 5.3   Relationship Matrix within an Organization

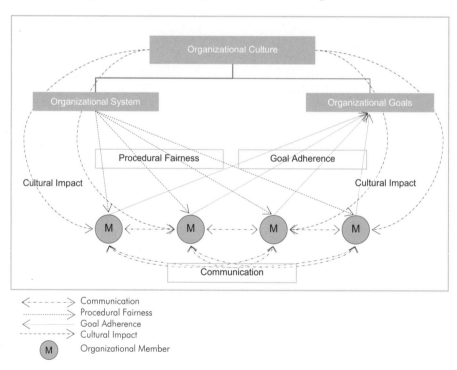

Organizational culture conceptualizes an organization's fundamental beliefs about important tasks and is depicted by the way preferred by an organization to function (Elangovan and Shapiro 1998). It represents a system of shared meaning of each organization that contains distinguishable characterization of innovation and risk taking; attention to details; outcome, people, and team orientation; aggressiveness; and stability (Crant 2000). The impact of organizational culture, as suggested, not only conveys the identity of the organization members, but also facilitates the generation of commitment. It cares about individual self-interest and thus enhances social system stability. In addition, organizational culture can function as a sense-making and control mechanism that guides and shapes the attitudes and behaviors of employees. Communication is a course of action in which an individual expresses meaning to another in order to effect information exchange through the use of symbols, verbally and/or non-verbally, consciously or unconsciously but intentionally. When two or more individuals come together, communication is unavoidable in order to achieve certain objectives. Communication connects people. It develops and improves through practices and is vital in social life. Moreover, this is also affected by personality and other personal factors. In terms of establishing working relationship, the impression of their initial interaction is critical for future relationship development. Whitener *et al.* (1998) suggested that seamless communications improves the quality of the relationship. In particular, if the first interaction is unimpressive, the parties may refrain from subsequent interactions, i.e., their relationship is implausible to make improvement.

Procedural fairness is considered as a good attribute of organizational culture. It is essential in fostering commitment. Equity underpins an individual's confidence, performance, efficiency, and attitude towards others (Das and Teng 2004). Experience of unfair treatments suppresses an individual's constructive behavior. Procedural fairness is likely to be reflected by company policies and personnel issues of an organization, i.e., the organizational system. Organizational system provides guidelines for members to behave and conduct business. A respectable organizational system is capable to balance the concerns between individualism and collectivism. Not only individual rights, self respect, and personal rewards and careers are to be taken into account, but also group activities and harmony of the unit are essential to the system.

Adherence to organizational goals is perceived as an individual's willingness to invest personal effort for the organization (Cook and Wall 1980). If an individual has a reasonable level of belief in the organization, he will be keen to commit himself. Leadership, motivation, and satisfaction are crucial for goal accomplishment. Leadership denotes the ability to influence a group towards the achievement of goals. Effective leadership style should be supportive, directive, participative, and achievement-oriented. It raises the overall energy or activity level and stimulates work-oriented motivation of organization members. Motivation denotes stimulating, directing, and sustaining employee efforts; it involves leading people to specific, goal-directed ways. The importance of people's expectations and rewards systems should not be neglected since they are essential to enrich and improve job performance. Job satisfaction refers to an individual's general attitude towards his job. Level of satisfaction depends on an organization member's feelings about his work, remuneration, promotion opportunities, supervision, co-workers, and working conditions. Having job satisfaction is vital in inducing organization citizenship behavior and positive attitude towards working parties (McAllister 1995).

5.4.2 The Elements of Trust

The identification of constructive characteristics of organizational behaviors penetrates the vivacity of trust in a healthy organizational relationship. Trust influences the effectiveness and performance of the work group because of its role in fostering organizational culture, communication, procedural fairness, and attachment to organizational goals. Before examining the contribution of trust in organizational relationship, it is essential to identify the basic ingredients of trust at this stage.

The exploration begins with the definitions of trust. Trust is an important concept for understanding interpersonal and group relationship in connection with individual behavior, managerial decisions, economic exchanges and other social issues. In the early years, researchers affiliated trust with expectation. For example, Deutsch (1958) described trust as the development of an individual's expectation on some motivational consequences. It is decided on the dependability of other individual or group. Golembiewski and McConkie (1975) opined that trust reflects an expectation about outcomes based on perceptions and life experiences. Trusting people expects satisfaction and fulfillment from others. Trust can also be derived from expectation of benevolent, constructive and advantageous response of a person to others (Mayer *et al.* 1995; Whitener *et al.* 1998). A number of researchers identified trust as a state of confidence. Deutsch (1973) and Scanzoni (1979) felt that one trusts because he has confidence in other parties' kindness rather than bad traits, and that other parties would provide some fulfillment to him. Zucker (1987) expressed similar points of view that trust builds on the confidence that a trust initiator will not be taken advantage of.

Willingness is another frequently used expression for trust. Scanzoni (1979) held an optimistic view that trust is manifested by the extent of one's willingness to collaborate with others. Trust is displayed if an individual is prepared and willing to submit to the threat of other parties betraying him (Whitener *et al.* 1998). The opinion of Mayer *et al.* (1995) is a combination of Scanzoni (1979) and Whitener *et al.* (1998); trust is indicated by one's willingness to look forward to attain satisfying outcome from others as well as his willingness to subject to risk of disappointment at the same time. Trust is the willingness to engage in a transaction in the absence of adequate safeguards to blend in the world of economics.

Trust is also related to positive belief. Tomkins (2001) found that trust carries a presumption of belief among people; a belief of the other parties' kindness (Whitener *et al.* 1998). Cumming and Bromiley (1996) expressed that trust is the belief that the parties involved will exercise good faith and unprejudiced effort. Moreover, Robinson (1996) related an individual's trust with other parties' beneficial and favorable actions towards him.

Other researchers identify trust as an outcome of behavior. For instance, Deutsch (1960) claimed that the derivation of trust depends on the behavior of others. Cumming and Bromiley (1996) pointed out that good-faith behavior is manifestation of trust while Whitener *et al.* (1998) argued that trust is cultivated through consistent behavior. Some "good" behaviors include acting with integrity, communicating openly, promptly and accurately, and working fairly (Das and Teng 1998; Whitener *et al.* 1998; Caldwell and Clapham 2003).

"Uncertainty" and "risk" are contextual ingredients of trust. Prima facie, there is no sign that uncertainty and risk are positive constructs to trust; however, their existence underlines the importance of trust. Golembiewski and McConkie (1975) purported that trust indicates some degree of uncertainty and implies something is put at risk. Bhattacharya *et al.* (1998) describe trust as the critical attitude when facing uncertain and risky environment. Das and Teng (1998) claimed that the building of trust stems from risk taking. It is the process of accepting risks which comprise "indiscretion, unreliability, poor coordination, cheating, abuse, neglect, self-esteem and misanticipation" (Sheppard and Sherman 1998). Apart from the above characterization, there are other expressions suggesting trust is something about reliance, hopefulness, optimism, honesty, mutuality, dependency, sharing of values, reciprocity, commitment, caring, and responsibility.

5.4.3 The Underpinnings of Trust

The preceding sections outline the various identifications of trust. In an attempt to conceptualize these findings, a diagrammatic presentation for construction of trust is given in Figure 5.4. Uncertainty and risk are the key trust inducers. Faced with these conditions, one need to derive trust among each other so that collective effort can be effected to overcome the difficulties ahead. Trusting behavior cannot be instigated without the willingness and confidence of the trust initiator. Nonetheless, the continuation of the trusting cycle depends very much on the reaction of the other parties involved because the trust initiator would expect reciprocal trusting behavior. This expectation if materialized would eventually become established as a belief that guides the behavior of the parties in subsequent dealings.

134

Figure 5.4   Construction of Trust

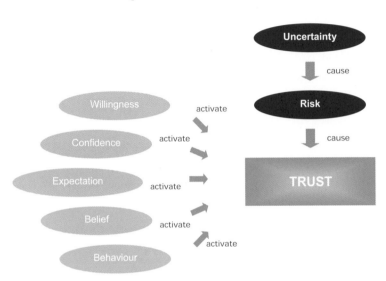

Uncertainty

The need to trust originates from uncertainty. In a state of uncertainty, an individual is not having sufficient information to map out a response that he has sufficient confidence. When encountering ambiguous circumstances, people may feel frustrated or difficult to overcome the problem without any support. Uncertainty may undermine success despite efforts of the parties involved. If this happens, disappointment will be resulted. The presence of trust may simplify the issues at stake thus enabling the derivation of viable solutions. Trust is present when there is an indistinct course of action in the future, i.e., trust is a way to bring solutions. Simmel (2004) commented that trust is "somewhere between total knowledge and complete ignorance". The need for trust is low when complete information is available. Except an individual is provided with sufficient information, the other option available for him is to predict the outcome in view of the uncertainties and risk involved. As such, the level of information available shall act as an indicator of the level of the need to trust.

Risk

Uncertainty about a situation induces risk. Risk carries the possibility of suffering loss, damage, or any other undesirable event. In a construction project, negative impacts include diminished quality of the end product, increased costs, delayed completion, or failure. Individuals should respond to the risk instead of ignoring it. Ignoring risk may leave the problem unresolved and at the end does not give any positive impact. An individual who prepares to take risk acknowledges the uncertainties involved. The statement pinpoints the importance of trust under ambiguous and risky tasks. An individual's intention to trust is shown by how he relies on the other party under precarious conditions. Trust is the acceptance of the risks (Sheppard and Sherman 1998). These risks are wide-ranging and while these are originated from an individual's psychological state, these risks are often difficult for others, whom they have a completely different mentality, to deal with. Therefore, to manage risk effectively an individual should be willing to trust. In other words, trust is an antidote to entail risk and indicates an individual's willingness to take risk.

The experience on uncertainty and risks stimulate the potential of trust development. These types of environment are ideal for trust to grow. Trust neutralizes the negative impact arising from uncertainties and risks. Trust in most cases acts as a catalyst to positivism; for instance, trust eliminates adversarial relationship, enhances better performance and communal relationship, improves working relationship, and sustains individual and organizational effectiveness.

Willingness

Trust is motivated by the level of an individual's willingness to establish affirmative relationships with other parties or organizations (Mayer *et al.* 1995). An individual's willingness to trust and act genuinely to another party can be built on this individual's confidence on the other party's anticipated fulfillment. Moreover, it is upon an individual's acceptability towards the risk he assumes and the degree of control he is willing to give up in return for his expected remuneration (Caldwell and Clapham 2003). The level of an individual's willingness to trust therefore varies with the potential of establishing business relationship with another party in an ambiguous environment; it also depends on his expectation on the others who will return constructive performance.

The above suggests that whether an individual will trust others depends on his expectation on the others in complying the "rule" of reciprocity. When an individual places great expectations on another party, simultaneously he is exposing himself. Facing such situation an individual shall have to make a conscious decision himself of whether to accept or repudiate. The decision making process reflects the degree of this individual's willingness to bear the risk. There is room for trust to develop only when this individual is willing to face up the risky situation. It is virtually impossible for an individual to become trusting in any way if he does not wish to take risk. An individual initiates to trust the other party is based on his confidence or expectation, i.e., there is no promise between the parties; even there is promise, risk is present and such promise may be fraudulent. Therefore, an individual has to make a decision whether he wants to trust the other party or not. Willingness is a decisive trigger; if an individual chooses to move forward, trust starts to grow; if an individual chooses to hold back, the trust cycle stops there.

Expectation

An individual who has expectation on another party suggests that he intents to trust that party. Rotter (1967) defined trust as a generalized expectation that someone is reliable in verbal statements. Yet, mere expectations are insufficient to induce trust of an individual. He believed that over generalized expectations are meaningless. More significantly, trust of an individual should be expressed coherently or with intentions. It should be grounded on expectations that are specific to that individual. In addition, Whitener *et al.* (1998) furthered the definition of trust that it should also indicate the expectancy on an individual about other parties' behavior or spoken or written speech which is kind and non-adversarial. Expectations of people vary from their perceptions and life experiences. An individual's expectations about trustworthy behavior tend to change as experience accumulates. Therefore, interaction is apparently important to fulfill one's expectations. People's observations and understandings influence their trusting attitude. They observe while communicating with others. Their observations on behaviors, oral or written, enhance their understandings on other people and thus build up their expectations on them. These expectations are influential to trust and they bring different conclusions to a trusting individual. The trusting individual will experience satisfying consequences if his expectation is fulfilled. However, dissatisfying consequences will be encountered if the trusting individual's expectation is not fulfilled.

Confidence

Confidence is another important trust motivator. From a social science perspective, trust is classified as mutual confidence (Barney and Hansen 1994); in particular, a trusting relationship is successful only if all the trusting parties have confidence in each other's. If confidence is unilateral, the equilibrium will be lost; ultimately, the trusting relationship ruptures. Nevertheless, a trusting individual is having confidence in other parties when they share commonalities in their value system, perceptive views and exposures (Wong *et al.* 2000). Trust has also been portrayed as an expression of confidence that the other side would not harm or put the trusting party at risk (Zucker 1987). The supporting rationale is quite straight forward. It is impossible to identify whether someone will treat you good or do you harm. Confidence helps individuals to think positively. Confronting tough situations together boosts mutual confidence and encourages people behave trustingly.

Belief

Trust is the belief on an individual's truthful and honest intention to perform for both parties' advantages. It is either an individual's belief or a group of individuals' common belief (Cummings and Bromiley 1996). An individual, who is willing to trust, represents that he believes the other party is reliable and he will not be disappointed by the other party. A trusting individual believes that the other party will not lie nor take advantages of him. Furthermore, he will not act against the other party's interests. These set of beliefs is unreserved. If the individual feels that the other party is not reliable or is taking advantage of him, his trusting belief will faint. Hardin (1991) suggested that reliance or belief is some kind of personal experience between individuals, with or without incentives, according to that individual's "experience". It is further suggested that belief is stronger if this is developed from early ages. Analogously, trusting experience during early stage of a project would be particularly reinforcing. However, McKnight *et al.* (1998) were in the view that if some contextual conditions, such as promises, contracts, regulations and/or guarantees, are in place, positive effect on belief building can be expected.

Interest and regularities are two other prerequisites for belief development (McKnight *et al.* 1998). In support of this proposition, Barney and Hansen (1994) noted that internal reward and compensation system should also be provided in order to build up an organization's values and beliefs. Tomkins (2001), in analyzing trust in construction contracting, suggested that provision of full information with plans, processes and anticipated results foster a trusting working environment. In general, once an individual's belief is strengthened, his trust level is also sustained.

Behavior

Luhmann (1979) suggested that trust can also be based on grounded expectations of another person's or institution's behavior. It could be expressed in behavior. Behavior denotes the way an individual performs. Attentions should be paid on "a particular way" when describing trusting behavior; as behaving in this particular way is the critical point for an individual to trust or not to trust. It is alleged trusting behavior is fairly consistent and integrated. The reason is obvious as consistent and integrated behavior would lead to higher predictability and thus confidence between the interacting parties intensifies.

Relationship between trust and behavior can also be demonstrated from the perspectives of predictability and confidence. One party's consistent behavior enhances predictability. Every individual has his own style. When there is one party approaching an individual, after several interactions, the party will become familiar with this individual and he could probably foresee and know how this individual would behave in the next occasion. Moreover, repeated dealing also produces high predictability. With increased predictability, the parties will have greater mutual confidence.

Not all kinds of behavior are constructive for trust development. Behaviors that induce trusting response include honest communication, providing accurate information, keeping promises, telling truths, sharing expectations, showing consideration and interest, protecting the parties' interests, etc. As soon as an individual shows favorable behaviors towards the other party, it is likely that the other party will behave in a constructive ways and avoid harming him. Consequently, common values and norms are developed between both parties and a satisfactory relationship is built, thus enabling a trusting environment to intensify.

5.4.4 The Contributions of Trust towards Organizational Relationship

After identifying the underpinning factors of trust, the contribution of trust towards organizational relationship is discussed. Trust reinforces individuals' affirmative willingness, confidence, expectation, belief, and behavior to overcome risk and uncertainty towards other individuals or organizations. The central role of trust can play is to establish and sustain co-operation.

Poor communication, ineffective leadership, and/or absence of common goals prevent teams from achieving their goals. How can trust help to improve? By providing objective information about motivating needs and styles of organization members, it helps to bridge gaps, establish faith, maximize the strengths of each of the organization member, and ultimately apply those strengths to achieve common goals. As a result, the primary procedure for resolution of unpleasant organizational relationship is to improve the culture of the organization. Individuals understand an organization through their exposure to organizational culture. Well developed organizational culture creates solidarity and enhances commitment. Whitener *et al.* (1998) recommend that trust is the key to a decent organizational culture. The presence of trust in corporations can speed up the development of organizational culture. Fair operating system, a balanced delegation of operating responsibility, an equitable resource allocation and control system, an affirmative human resource and performance evaluation policies, together

140

with a devotion to set up common goals and mutual dependence are catalytic towards a trusting culture. It is believed that a good organizational culture provides confidence for individuals to behave well for the organization. Such background prompts their willingness to believe in the organization and comply with organizational policy and procedures in order to create a sincere working environment.

Trust enhances good communication. Seamless communication reflects an individual's positive willingness, confidence, expectation, belief, and behavior that link people to get closer with each other. The two-way communication enables each party gaining better knowledge from the other party. Even if the communication process involves conflicts, they may be constructive to the functioning of a group. Enhanced communication gets people closer together. High-quality communication reduces uncertainty. Trust increases the probability for people to share information and better understanding of each other. Performance would improve if members of an organization are considerate and care about other's concern. The likelihood of acting dishonestly would be diminished.

Perceived organizational fairness renders individuals' comfort over organizational decisions. Regarding organization policies, it is preferable for all parties to have maximum opportunity to participate in processes and systems that are fair. In response to an equitable organization environment, individuals intend to trust and become loyal and respectful to their company.

A trusting attitude is very important for individuals to submit to the organization. They are more faithful and their behaviors will be in line with organizational goals. Individuals having high level of commitment to organizational goals are more willing to engage in organizational citizenship behavior. With higher motivation, more satisfying performance can be expected. In other words, an enthusiastic staff force who loves their jobs, honors their leaders, commits to their organizations is what every organization looks for.

5.5 Summary

This chapter reviews some behavioral aspects in organizations, notably organizational culture, communication, procedural fairness, and attachment to organizational goals. These are reliable tools in expressing organizational relationship. Effecting favorable organizational relationship requires an effective instrument. Previous literatures showed that trust is one of the answers to improve organizational relationship. This chapter, through a case study, highlights first the importance of behavioral change in implementing co-operative contracting. The inducers of trust are then deliberated. Uncertainty and risk are the two important trust inducers. These two unfavorable conditions cause anxiety. Trust is the natural cure if collective efforts can be synergized to respond to the uncertainty and risk. Nonetheless, trust needs to be nutured. Willingness, expectation, confidence, belief, and behavior are identified as the five trust generators. They enhance the trust level between individuals. The presence of trust also improves organizational culture that drives good behaviors of individual organization members. Communication with trusting intention also promotes honest sharing between individuals. Fair procedures in organization derive common objectives and attachment to organizational goals. It is suggested that the contribution of trust in organizational relationship is analogous the relationship between clients/consultants and contractors in the construction industry as construction projects have to be accomplished through co-operations among organizations.

References

Associated General Contractors of America. 1991. *Partnering: A Concept for Success.* Associated General Contractors of America.

Barney, J. B., and M. H. Hansen. 1994. Trustworthiness as a source of competitive advantage. *Strategic Management Journal* 15(special issue): 175–190.

Bayliss, R. 2002. Project partnering—A case study on MTRC Corporation Ltd's Tseung Kwan O Extension. *HKIE Transactions* 9 (1): 1–6.

Bayliss, R., S. O. Cheung, C. H. Suen, and S. P. Wong. 2004. Effective partnering tools in construction: A case study on MTRC TKE Contract 604 in Hong Kong. *International Journal of Project Management* 22(3): 253–263.

Beccera, M., and L. Huemer. 2000. Moral character and relationship effectiveness: An empirical investigation of trust within organizations. In *Proceedings of 2nd ISBEE World Congress, Business, Economics and Ethics, Sao Paulo, July 19–23.*

Bernstein, C., R. J. Emerson, and A. Gabor. 1989. A paper on trust. http://hamp.hampshire.edu/~AWAKE/papers/891957223.html [accessed June 2006]

Bhattacharya, R., T. M. Devinney, and M. M. Pillutla. 1998. A formal model of trust based on outcomes. *The Academy of Management Review* 23(3): 473–490.

Black, C., A. Akintoye, and E. Fitegerald. 1999. An analysis of success factors and benefits of partnering in construction. *International Journal of Project Management* 18(6): 423–434.

Brenkert, G. 1998. Trust, business and business ethics: An introduction. *Business Ethics Quarterly* 8(2): 195–203.

Caldwell, C., and S. E. Clapham. 2003. Organizational trustworthiness: An international perspective. *Journal of Business Ethics* 47(4): 349–364.

Carnevale, D. G., and B. Wechsler. 1992. Trust in the public sector: Individual and organizational determinants. *Administration & Society* 23(4): 471–494.

Cook, J., and T. Wall. 1980. New work attitude measures of trust, organizational commitment and personal need non-fulfillment. *Journal of Occupational Psychology* 53(1): 39–52.

Construction Industry Institute Australia (CIIA). 1996. *Partnering: Models for Success. Research Report No. 8.* Brisbane, Australia.

Construction Industry Review Committee. 2001. *Tang's Report on the Hong Kong Construction Industry Reform.*

Crant, M. J. 2000. Organizational behavior. http://www.nd.edu/~mba/thebook/Org_Behavior.htm#Resources [accessed June 2006]

Cummings, L. L., and P. Bromiley. 1996. The Organizational Trust Inventory (OTI): Development and validation." In *Trust in Organizations: Frontiers of Theory and Research,* ed. R. M. Kramer and T. R. Tyler, 302–330. SAGE Publications.

Das, T. K., and B. S. Teng. 1998. Between trust and control: Developing confidence in partner cooperation in alliances. *The Academy of Management Review* 23(3): 491–512.

Das, T. K., and B. S. Teng. 2004. The risk-based view of trust: A conceptual framework. *Journal of Business and Psychology* 19(1): 85–116.

Deutsch, M. 1958. Trust and suspicion. *Journal of Conflict Resolution* 2(4): 265–279.

Deutsch, M. 1960. Trust, trustworthiness and the F-scale. *Journal of Abnormal and Social Psychology* 61: 138–140.

Deutsch, M. 1973. *The Resolution of Conflict: Constructive and Destructive Process.* New Haven: Yale University Press.

DeVilbiss, C. E., and P. Leonard. 2000. Partnering is the foundation of a learning organization. *Journal of Management in Engineering* 16(4): 47–57.

Doney, P. M., J. P. Cannon, and M. R. Mullen. 1998. Understanding the influence of national culture on development of trust. *The Academy of Management Review* 23(3): 601–620.

Elangovan, A. R., and D. L. Shapiro. 1998. Betrayal of trust in organizations. *The Academy of Management Review* 23(3): 459–472.

Ford, D. 2002. Trust and knowledge management: The seeds of success. Working paper wp01-08. Queens' KBE Centre for Knowledge-based Enterprises, Queens University.

Golembiewski, R.T., and M. McConkie. 1975. The centrality of interpersonal trust in group process. In *Theories of Group Processes,* ed. C. L. Cooper, 131–185. London: Wiley.

Hardin, R. 1991. Trusting persons; trusting institutions. In *Strategy and Choice,* ed. R. J. Zeckhauser, 185–209. Cambridge, Mass.: MIT Press.

Hart, K. M. 1988. A requisite for employee trust: Leadership. *Psychology: A Journal of Human Behavior* 25: 1–7.

Hawke, M. 1994. Mythology and reality—The perpetuation of mistrust in the building industry. *Construction Papers of the Chartered Institute of Building* 4.

Jannadia, M. O., S. Assaf, A. A. Bubshait, and A. Naji. 2000. "Contractual Methods for Dispute Avoidance and Resolution (DAR)." *International Journal of Project Management 2000* 18(6): 41–49.

Jones, G., and J. George. 1998. The experience and evolution of trust: Implications for cooperation and teamwork. *Academy of Management Review* 23(3): 531–548.

Kumaraswamy, M. M., and J. D. Matthews. 2000. Improved subcontractor selection employing partnering principles. *Journal of Management in Engineering* 16(3): 47–57.

Larson, E., and J. A. Dexler. 1997. Barriers to project partnering: Report from the firing line. *Professional Journal of Project Management Institute* 28(1): 46–52.

Law, F. 2004. Partnering—The client's perspective. In *CII-HK Conference 2004 on Construction Partnering, Hong Kong*.

Lewis, J. D., and A. Weigert. 1985. Trust as a social reality. *Social Forces* 63: 967–985.

Li, H., W. L. Cheng, and P. Love. 2000. Partnering research in construction, engineering. *Construction and Architectural Management* 7(1): 76–92.

Luhmann, N. 1979. *Trust and Power*. New York: Wiley.

Matthai, J. M. 1989. Employee perceptions of trust, satisfaction and commitment as predictors of turnover intentions in a mental health setting. Doctoral diss., Prebody College and Vanderbilt University. Dissertation Abstracts International, DAI-B51/02.

Matthews, J. D. 1996. A project partnering approach to the main contractor-subcontractor relationship. PhD thesis, Loughborough University, Loughborough, UK.

Mayer, R. C., J. H. Davis, and F. D. Schoorman. 1995. An integrative model of organizational trust. *The Academy of Management Review* 20(3): 709–734.

McAllister, D. J. 1995. Affect- and cognition-based trust as foundations for interpersonal cooperation in organizations. *Academy of Management Journal* 38(1): 24–59.

McDermott, P., M. M. A. Khalfan, and W. Swan. 2005. "Trust" in Construction Projects. In *Proceedings of International Commercial Management Symposium, University of Manchester, 7 April 2005*. p. 97.

McKnight, D. H., L. L. Cumming, and N. L. Chervany. 1998. Initial trust formation in new organizational relationships. *The Academy of Management Review* 23(3): 473–490.

Moore, M. 1999. *Commercial Relationship*. UK: Tudor Business Publishing Limited.

Noorderhaven, N. 1999. National culture and the development of trust: The need for more data and less theory. *Academy of Management Review* 24(1): 9–10.

Nooteboom, B. 1999. *Inter-firm Alliances—Analysis and Design*. London: Routledge.

Nyhan, R. C., and H. A. Marlowe, Jr. 1997. Development and psychometric properties of the organizational trust inventory. *Evaluation Review* 21(5): 614–635.

Pennings, J. M., and J. Woiceshyn. 1987. A typology of organizational control and its metaphors. In *Research in the Sociology of Organizations* 5, ed. S. B. Bacharach & S. M. Mitchell. 75–104.

Piper, B. J. 2001. Partnering: a dream? *Newsletter of the Hong Kong Institute of Surveyors* 8(10): 22–23.

Rapoport, A., A. M. Chammah, and A. Arbo. 1965. *Prisoner's Delimma*. USA: University of Michigan Press.

Reading Construction Forum. 1995. *Trusting the Team: The Best Practice Guide to Partnering in Construction.* Centre for Strategic Studies in Construction, Reading, UK.

Robinson, S. L. 1996. Trust and breach of the psychological contract. *Administrative Science Quarterly* 41(4): 574–599.

Rotter, J. B. 1967. A new scale for the measurement of interpersonal trust. *Journal of Personality* 35(4): 651–665.

Rousseau, D., S. Sitkin, R. Burt, and C. Camerer. 1998. Introduction to special topic forum. Not so different after all: A cross-discipline view of trust. *Academy of Management Review* 23(3): 393–404.

Sanders, S. R., and M. M. Moore. 1992. Perceptions of partnering in the public sector. *Project Management Journal* 22(4): 13–19.

Scanzoni, J. 1979. Social exchange and behavioral interdependence. In *Social Exchange in Developing Relationships,* ed. R. L. Burgess and T. L. Huston, 61–98. New York: Academic Press.

Sheppard, B. H., and D. M. Sherman. 1998. The grammars of trust: A model and general implications. *The Academy of Management Review* 23(3): 422–437.

Simmel, G. 2004. *The Philosophy of Money.* 3rd ed. London and New York: Routledge.

Tomkins, C. 2001. Interdependencies, trust and information in relationships, alliances and networks. *Accounting, Organizations and Society* 26(2): 161–191.

Whitener, E. M., S. E. Brodt, M. A. Korsgaard, and J. M. Werner. 1998. Managers as initiators of trust: An exchange relationship framework for understanding managerial trustworthy behavior. *The Academy of Management Review* 23(3): 513–530.

Whitney, J. 1999. *The Trust Factor, Liberating Profits and Restoring Corporate Vitality.* UK: McGraw Hill.

Wong, E. S., D. Then, and M. Skitmore. 2000. Antecedents of trust in intra-organizational relationships within three Singapore public sector construction project management agencies. *Construction Management and Economics* 18(7): 797–806.

Wood, G., and P. McDermott. 1999. Searching for trust in the UK construction: An interim view. In *CIB W92 International Procurement Systems Conference, Thailand.*

Zucker, L. G. 1987. Institutional theories of organization. *Annual Review of Sociology* 13(1): 443–464.

Trust Factors in Co-operative Contracting: Views of Parties of a Partnering Dance

Sai On Cheung
Peter Shek Pui Wong

Acknowledgements

Part of the content of this chapter has been published in Vol. 22 of the *International Journal of Project Management*. The authors thank the publisher for the permission to use the content there-in in this Chapter.

6.1 Introduction

Trust is one of the most essential synthesizers of interpersonal relationships. It is also widely accepted that a high level of trust is essential to organizational effectiveness, particularly in an industry as culturally diverse as the construction industry (Good 1988).

It has been pointed out by RAPS (2000): "Although an organization obviously cannot succeed without high levels of trust among members, most organizations do little to actively build trust". Fairholm (1994) suggested that "Trust has not been given much attention by either academics or practicing professionals". Based on the literature reviews, he found that: "Under conditions of high trust, problem solving tends to be creative and productive. Under conditions of low trust, problem solving tends to be degenerative and ineffective." Kumar (1996) suggested that trust "Creates a reservoir of goodwill that helps preserve the relationship when, as will inevitably happen, one party engages in an act that its partner considers destructive."

These findings are inspirational in enhancing the understanding of the roles of trust in enabling organizational improvement through relationship building. However, it has also been found in these studies that as new teams are organized, the issue of trust is rarely considered and addressed at a level that it deserves (Hawke 1994, Zaghloul and Hartman 2000). Creating trusting relationships is by no means a simple process, particularly in the construction industry. There are challenges that need to be overcome in order to build trust effectively in this industry.

The construction industry has placed strong faith in partnering to achieve cost effectiveness, work efficiency, opportunities for innovations, equitable risk allocation and less confrontation (CII 1989, CIRC 2001, Black *et al.* 2000, Bayliss 2002). For partnering to succeed, Black *et al.* (2000) pointed out that developing trust among partners is the most important factor. However, they also indicated that "few industries suffer more from conflict than construction." As such, Hawke (1994) argued that building mutual trust in construction is a myth. She opined that "trust within the building industry has not deteriorated nor diminished. Nonetheless, mistrust in construction contracting is deep-seated and long-standing. The myth is that the construction industry has always been fragmented and contentious, and therefore, it always will be." How can trust be developed and maintained in projects that are relationship based? This chapter reports a study designed to identify key trust factors from the perspectives of two groups: a) clients and consultants and b) contracting organizations.

6.2 Trust Attributes in Co-operative Contracting

Trust is a complex construct with multiple bases, levels and determinants (Rousseau *et al.* 1998). It is dynamic and either growing or diminishing (Hawke 1994). It is often associated with situations involving personal conflict, outcome uncertainty and problem solving. It is a prediction and expectation of future events. Varying in intensity, this is the confidence in and reliance upon the prediction (Nyhan *et al.* 1997).

In fact, trust has been expanded to a variety of theories and concepts applied in different fields according to their natures and characteristics (Frost *et al.* 1978, Good 1988, Jones and George 1998, Ford 2001). Nevertheless, considering the period of collaboration, the project complexity, and the changing project conditions, the theories and concepts of trust in construction partnering are different from the other settings (Rosenfeld *et al.* 1991, Kramer and Tyler 1996).

Hartman (2003) identified three bases of trust that explain why people place their trust on another party in construction projects. These are Competence Trust, Integrity Trust and Intuitive Trust. Competence Trust is based on the perception of others' ability to perform the required work. Partners' Competence Trust can be gained by observable proofs like track record, experience or connections with professional bodies (Romahn and Hartman 1999, Zaghloul and Hartman 2000). Integrity Trust (or Ethical Trust) is based on the perception of others' willingness to protect the interest of their counter parts over the construction project. The level of integrity trust is highly affected by the values, morals, ethics and cultural backgrounds of the parties. Generally, establishing open communication is critical to enhance or gain partners' Integrity Trust.

Intuitive Trust (or Emotional Trust) is founded upon the party's prejudice, biases or other personal feelings towards the counter parts. Intuitive Trust is the perception which is hardly affected by the instant performance of the parties but the long-term relationships among them.

The conceptualization of trust by Hartman (2003) has put trust in construction in perspective. Nonetheless, Hartman (2003) also reminds the need to watch out for cultural and geographical differences as suggested by Romahn and Hartman (1999). For example, many of the research in trust is done in North America. Hence, attributes of trustworthiness are inevitably defined and measured based on North American values, ethics and morals (Romahn and Hartman 1999). The study reported in this chapter was conducted in Hong Kong and may bring out the cultural dimension when compared with Hartman's (2003) conceptualization. In addition, partners in a construction project may rely on different types of trust to enhance their trust levels. In view of the cyclical and dynamic nature of trust, it is crucial to study which type of trust is most effective with respect to the parties of the partnering dance.

In this chapter, the situations/circumstances where trust can be developed are described as trust attributes. The related work of the Centre for Construction Innovation in the UK provides a good list of twelve trust variables obtained from literatures and interviews with industry practitioners. In addition, Fukuyama (1996) and Kadefors (2004) suggested that the use of alternative dispute resolution and satisfaction on the contract terms are also trust attributes in construction partnering. As a result, 14 trust attributes were used in this study. Their descriptions and references were presented in Table 6.1.

150

Table 6.1   14 Trust Attributes for Construction Partnering

Trust Attributes	Descriptions	Justification
Competence of Work (Competent)	Project team members will trust each other if both of their "behaviors" and "outcomes" are competent [William et al. 2002].	Undertaking interviews with the industry experts from different disciplines and levels in construction industry Centre for Construction Innovation and the Salford Centre for Research and Innovation in U.K
Problem Solving Ability (Problem Solving)	Construction personnel saw problem solving as an important element in building trust, especially if it is solved at the early stage [USCCI 2002].	
Frequency and Effectiveness of Communication (Communication)	Open and frequent communication and maintaining open-door policies to each other results from a willingness of the partners to create transparency in the relationship [Sarker et al. 1998, Wood et al. 2001].	
Openness and Integrity of Communication (Openness)	Failure of integrity involves lying; cheating or hiding facts happened in the project team will tarnish trust [Rosenfeld et al. 1991, Sarker et al. 1998].	
Alignment of Effort and Rewards (Alignment)	Benefits received should be fair and match with the input efforts or mistrust will result [Wood et al. 1999].	
Effective and Sufficient Information Flow (Information Flow)	Effective and sufficient information flow would reduce risk and uncertainty of the work [USCCI 2002].	
Sense of Unity (Unity)	Trust can be built by understanding and appreciating partner's requirements and difficulties and looking to meet partner's expectation [USCCI 2002]."	
Respect and Appreciation of the System (Respect)	A respect on the mutual dependence project management system is a source of trust [Wood et al. 2001, Gill and Bulter 1996].	
Compatibility (Compatibility)	Project team members will trust each other when they share similar cultures and values [Sarker et al. 1998, Wood et al. 1999].	
Long-term Relationships (Long-term Relations)	Long term relationships among partners will lead to trust [Morgan and Hunt 1994, USCCI 2002].	
Financial Stability (Financial)	The financial status of the company affects the decision to trust. Contractors who have a healthy financial status are trustworthy in views of the clients as their risks to make profits by finding loopholes in contract or applying unreasonable claims are lowered down [USCCI 2002].	
Reputation (Reputation)	Companies with higher reputation are more trustworthy as they do not want to lose their valuable asset [Gambetta 1988, Wood et al. 2001].	
Adoption of ADR Techniques (Adopt ADR)	The implementation of ADR techniques before litigation as stated in the contract would also gain trust from other parties. These contracting parties will feel that their partners are willing to seek for win/win resolution sincerely without destroying the cooperation harmony [Cheung 1999].	Source of institutional-based trust in construction. (Fukuyama 1996, Kadefors 2004)
Satisfaction of Contract Terms and Agreements (Satisfactory Terms)	Equitable agreements or contracts terms can help the contracting parties to establish trust and sustain cooperation since their perceived benefits are secured [Bonet et al. 2000].	

6.3 Views of Trust Factors

In this section, a study designed to unveil the views of parties to a co-operative contracting is reported. Partnering projects were targeted because it is a form of construction project procurement approach that exemplifies co-operative contracting in construction. This notion has been discussed in detail in Chapter four.

As such, data on the importance of trust factors, project performance as well as trust level of project partners were collected through a project specific data collection questionnaire. A sample of questionnaire is given in Figure 6.1. This chapter shall report first on the evaluation of the importance of trust attributes. Chapter seven shall report the findings on the relationship between trust factors and partnering success. In addition an evaluation on who is the best trust initiator in partnering projects is also described.

6.3.1 Data Collection

To accomplish the research objectives, a postal questionnaire survey was performed for data collection. Part 2 of the questionnaire was designed to solicit the respondents' assessment on the degree of importance of the fourteen trust attributes in affecting partners' trust level by a seven-point Likert scale (1-not important, 7-very important). To ensure the relevance of the responses, questionnaires were sent to practitioners having experiences with partnering projects. The names and the postal addresses of the respondents are obtained from the web pages of the local professional institutes as well as the *Hong Kong Builder Directory*. In order to safeguard the reliability of the received responses, respondents were asked to provide information on their experiences in partnering projects. If the respondents replied that they had not taken part in any partnering projects, their returned questionnaires were discarded. Therefore, the reliability of the survey results was assured.

Figure 6.1 Sample of Questionnaire

Part 1 Personal Information				
Please fill in the information (or circle the appropriate choice) for each question				

Q1.1	Do you have experience to take part in partnering project(s)?	a) Yes	b) No	
Q1.2	Your role in this partnering project	a) Developer	b) Contractor	c) Consultant
Q1.3	Working experience	a) < 5 years	b) 5-10 years	c) 11-15 years
		d) 16-20 years	e) >20 years	

PART 2 The Trust Factors in Partnering Projects

		Very low					Very high	
Q2.a	The trust level on my partner	1	2	3	4	5	6	7

According to the answers you provided in Q2.a, please circle the no. that best reflects the degree of importance of the following 14 attributes in developing trust

		Very low					Very high	
Q2.1	The competence of work of my partner	1	2	3	4	5	6	7
Q2.2	Problem solving ability of my partner	1	2	3	4	5	6	7
Q2.3	The frequency and the effectiveness of communication of my partner	1	2	3	4	5	6	7
Q2.4	The openness and integrity of communication of my partners	1	2	3	4	5	6	7
Q2.5	Alignment of effort and rewards among partners	1	2	3	4	5	6	7
Q2.6	Effective and sufficient information sharing with my partner.	1	2	3	4	5	6	7
Q2.7	The sense of unity of my partner	1	2	3	4	5	6	7
Q2.8	Partners' respect, believe and rely on the project management system	1	2	3	4	5	6	7
Q2.9	The compatibility of my partner	1	2	3	4	5	6	7
Q2.10	A long-term relationships with my partner	1	2	3	4	5	6	7
Q2.11	The financial stability of my partner	1	2	3	4	5	6	7
Q2.12	Adoption of ADR techniques by my partner	1	2	3	4	5	6	7
Q2.13	The reputation of my partner	1	2	3	4	5	6	7
Q2.14	My partners' satisfaction on the contract terms and agreement	1	2	3	4	5	6	7

PART 3 Achievement of project goals

In considering the successfulness of the partnering project, please indicate the degree of importance of achieving following project goals:-

		Not important					Very Important	
TIME								
Q3.1	The project can meet the committed target date for completion as stated in the partnering charter	1	2	3	4	5	6	7

Continued on next page

Figure 6.1—*Continued*

Q3.2	Resolution of problems become more efficient	1	2	3	4	5	6	7
Q3.3	The work process throughout the project is smooth and efficient and the redundant work is eliminated	1	2	3	4	5	6	7
Q3.4	Stakeholders can make decisions efficiently in problem solving	1	2	3	4	5	6	7
COST								
Q3.5	The project can meet the committed target budget for completion as stated in the partnering charter	1	2	3	4	5	6	7
Q3.6	Cost saving proposals efficiently lead clients and contractors to achieving higher profit margins	1	2	3	4	5	6	7
Q3.7	Effective utilization of resources reduces wastage of both labor and materials	1	2	3	4	5	6	7
Q3.8	Reduced claims, variations and risk of litigation	1	2	3	4	5	6	7
QUALITY								
Q3.9	The project can meet the committed target quality as stated in the partnering charter	1	2	3	4	5	6	7
Q3.10	Continuous improvement of quality can be achieved by joint evaluation and implementation of innovative or effective methods	1	2	3	4	5	6	7
Q3.11	Continuous improvement of safety levels can be achieved by joint evaluation	1	2	3	4	5	6	7
Q3.12	Stakeholders will accept others' mistakes and consider remedying the fault together	1	2	3	4	5	6	7
COMMUNICATION								
Q3.13	The meetings are frequent, sufficient and effective	1	2	3	4	5	6	7
Q3.14	All stakeholders' interests are considered and respected during discussion	1	2	3	4	5	6	7
Q3.15	Problems are recognized and rectified at the earliest stage through meetings or other communication	1	2	3	4	5	6	7
Q3.16	Sharing of information is open, honest, accurate, timely, helpful and trust manner	1	2	3	4	5	6	7
MANAGEMENT								
Q3.17	A good leader chairs the partnering meeting, monitoring and supporting the functionality of partnering	1	2	3	4	5	6	7
Q3.18	All representatives involved are of sufficient caliber to make agreeable decisions	1	2	3	4	5	6	7
Q3.19	Partnering tools including problem identification, conflict escalation procedures, ADR approaches and an evaluation methodology.	1	2	3	4	5	6	7
Q3.20	Mutual goals can be achieved and mutual trust can be enhanced	1	2	3	4	5	6	7

6.3.2 The Response Rate

A total of 120 questionnaires were sent to private and public sector developers, consultant firms and contractor firms. The response rate of all three groups were higher than 30% and numbers of valid responses were considered reasonably good compared with similar studies conducted in the United States (Sarker *et al.* 1998) and Hong Kong (Cheng and Li 2001). The details of the above are summarized in Table 6.2.

Table 6.2   Questionnaire Sent, Received and Valid Responses

	Questionnaires Sent (No.)	Questionnaires Received (No. / %)	Valid Response (No. / %)
Developers (Public and Private)	43	15 (34.9%)	13 (30.2%)
Consultants	33	17 (51.5%)	14 (42.4%)
Contractors	44	33 (75.0%)	24 (54.5%)
Total	120	65 (54.2%)	51 (42.5%)

The enlisting of the 14 trust attributes helps to put the study on trust in partnering in perspective. However, to assist management to direct their efforts to foster partnering success, it is essential to identify first those trust attributes that are most effective in building trust. In this connection, Gill and Bulter (1996) argued that relying on contract terms is a feature hampering partners' trust. Morgan and Hunt (1994) emphasized that fostering long-term co-operation would enhance partners' trust. These studies were mainly anecdotal and a more systematic treatment would be useful. Hence, the first objective of this study is to identify the relative importance of the trust attributes in affecting the partners' trust level. Contracting involves principally two groups of participants: a) clients and consultants and b) contracting organizations. Although in partnering both groups should strive for common goals, however, with different backgrounds and business objectives, their views may be different. Therefore, the study results are arranged in two groups for comparison purposes.

The 14 trust attributes as identified in Table 6.1 are often discussed in different contexts. In an attempt to make the notion of trust in construction partnering more precise and amenable for analytical grip, the second objective of this study is to group and interpret these 14 attributes by a smaller number of factors. The statistical technique of Principal Component Factor Analysis (PCFA) was applied to condense the 14 trust attributes into a smaller and more manageable number of factors (Hair *et al*. 1998). The factors identified generally could better represent the underlying construct in a concise and interpretable form (Dulaimi *et al*. 2002).

As for the experience of the respondents, as shown in Figure 6.2, 62% of them have over 10 years of working experience. With this, it was considered that the data set should form a good base for the proposed analysis and the result thereof was reflective of the prevailing industry's opinion in Hong Kong.

Figure 6.2   Respondents' Working Experience

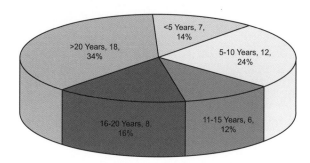

6.3.3 Discussion

Relative Importance Ranking of the 14 Trust Attributes

The importance of the 14 trust attributes were ranked by their mean scores derived from all valid responses. If two or more trust attributes happened to have the same score, the one with the lower standard deviation was assigned the higher ranking. In this research, the importance ranking results were separated into two groups: Clients & Consultants; and Contractors, as shown in Table 6.3.

Table 6.3   Relative Importance Ranking of the 14 Trust Attributes

Trust Attributes	Clients and Consultants			Contractors		
	Mean Score	Std. Dev.	Ranking	Mean Score	Std. Dev.	Ranking
Reputation	5.2174	1.2044	1	5.3214	0.8630	1
Satisfactory Terms	5.0435	1.1069	2	4.3929	1.2864	12
Openness	4.8261	1.2304	3	4.8929	1.0660	7
Information Flow	4.7391	1.1369	4	4.9286	1.0862	6
Alignment	4.6957	1.1455	5	5.1429	0.8483	2
Adopt ADR	4.6364	1.4325	6	5.1071	1.0306	4
Financial	4.6087	1.6717	7	4.6071	1.1001	11
Communication	4.5652	1.3425	8	5.1429	1.2683	3
Competent	4.4348	1.1995	9	5.0357	0.7926	5
Unity	4.3913	1.4691	10	4.8214	1.1239	9
Problem Solving	4.3913	1.2336	11	4.8571	1.1127	7
Respect	4.3043	1.1455	12	4.7143	0.8967	10
Long-term Relations	3.5652	1.5023	13	4.2143	1.7503	13
Compatibility	2.0476	1.5645	14	3.2143	1.7292	14

Referring to Table 6.3, trust attributes "Long Term Relations" (ranked 13th) and "Compatibility" (ranked 14th) have low mean scores. The results indicate that these two attributes are relatively less important in affecting partner's trust level in the view of the respondents of both the Contractors Group and the Clients and Consultants Group.

For the rest of the attributes, their mean scores lie in a very close range from 4.3 to 5.3. This indicates that the levels of importance of these attributes perceived by both groups of respondents are very close. This result pattern provides little information for management action and a small difference of their mean scores would cause a noticeable variation of the rankings. This explains why the ranking positions of these trust attributes from the two groups are quite different. In this respect, grouping of trust attributes into a smaller number of factors may prove useful. This is because the factors identified could better represent the underlying construct of the similar type of attributes in a more concise and interpretable form (Dulaimi *et al.* 2002). In addition, the findings could be critically compared with the Hartman's Trust Model (Hartman

2003) in order to discover any trust behavioral difference between construction partners in Hong Kong and North America.

Grouping Trust Attributes into Smaller Number of Factors by PCFA

Principal component factor analysis (PCFA) is an effective tool to factorize a large number of attributes into a smaller number of factors to enhance a more systematic and effective data interpretation (Hair *et al.* 1998, Cheung 1999). As such, in an attempt to investigate the underlying construct of the 14 trust attributes; PCFA was performed by the use of the statistical package of social science (SPSS). To achieve a simpler and pragmatically more meaningful factor solution, VARIMAX rotations were performed to enhance factor interpretations (Sharma 1996, Hair *et al.* 1998). The results of the factor analyses for the Clients and Consultants Group and the Contractors Group after VARIMAX rotation are summarized in Tables 6.4 and 6.5.

Clients and Consultants Group

For Clients and Consultants Group (Table 6.4 refers), both results from the Bartlett Test of Spericity and the KMO Measure of Sampling Adequacy (which is higher than the threshold of 0.500) indicated that the sample data from the Clients and Consultants Group was adequate for factor analysis (Sharma 1996, Hair *et al.* 1998, Cheung 1999).

Applying the Eigenvalue greater than 1 rule (Sharma 1996), four trust factors were identified in the Clients and Consultants Group after the VARIMAX rotation. The extracted factors explained 82.00% of the total variance, which was considered sufficient to explain trust using the extracted attributes. Five attributes were extracted as significant in Factor 1. They were Problem Solving, Competent, Unity, Communication and Respect. Referring to the nature of these attributes in Table 6.9, this factor is describing the working performance of partners. Therefore, Factor 1 was interpreted as Partners' Performance. Openness, Alignment, Financial Stability, Adopt ADR and Information Flow loaded highly in Factor 2. Collectively, Factor 2 can be interpreted as Partner's Permeability. Factor 3 consists of the attributes; Satisfactory Terms and Reputation. The attributes aligned in this factor are based upon the perceived reliability of the formalized systems rather than inter-organization trust. Thus, Factor 3 is interpreted as System-based Trust. Long-term Relations and Compatibility are the attributes extracted for Factor 4. Both attributes concern the humans' emotional connections with long-term cooperation (Bernstein *et al.* 1989). Thus, Factor 4 is interpreted as the Relational Bonding between Partners.

Table 6.4 Factor Analysis and Total Variance Explained—Trust (Clients and Consultants)

Trust Attributes	Factor			
	1	2	3	4
Problem Solving	.910	.235	3.328E-02	-9.564E-02
Competent	.861	1.552E-03	.359	.145
Unity	.845	.399	.235	-1.971E-02
Communication	.697	.462	-9.766E-02	.173
Respect	.662	.556	-.200	-3.597E-02
Openness	.153	.922	.183	-.141
Alignment	.305	.722	.326	-.230
Financial stability	.629	.694	-1.405E-04	4.099E-02
Adopt ADR	.339	.687	7.567E-02	.263
Information Flow	.243	.670	.194	.397
Satisfactory terms	8.409E-03	7.385E-02	.898	-.177
Reputation	.398	.447	.639	3.275E-02
Long-term Relations	8.359E-03	.195	-2.262E-02	.902
Compatibility	2.346E-02	-.324	-.447	.723
% Variance	28.963	27.569	13.083	12.397
Eigenvalue	4.055	3.860	1.832	1.733
Internal consistency reliability (Cronbach alpha)	.926	.887	.669	.657
Kaiser-Meyer-Olkin Measure of Sampling Adequacy	.700			
Bartlett Test of Spericity *Approx. Chi Square* *df* *sig*	224.985 91 .000			

Table 6.5   Factor Analysis and Total Variance Explained—Trust (Contractors)

Trust Attributes	Factor			
	1	2	3	4
Problem Solving	.895	9.513E-02	.129	1.416E-03
Competent	.876	-.125	5.344E-02	-.133
Unity	.802	2.850E-02	-.126	.276
Communication	.726	.404	.194	.118
Respect	.721	.306	-.195	.423
Openness	.674	.511	-3.574E-02	-.174
Alignment	.454	.348	-.428	.377
Financial stability	5.406E-02	.784	.296	.330
Adopt ADR	1.684E-02	.704	-7.678E-03	-3.040E-02
Information Flow	.557	.698	-.111	9.620E-02
Satisfactory terms	.206	-4.014E-02	.849	-.121
Reputation	.111	8.249E-02	.780	.372
Long-term Relations	.251	-.132	-.682	.209
Compatibility	2.830E-02	7.660E-02	-3.433E-02	.932
% Variance	31.076	16.379	15.652	11.744
Eigenvalue	4.351	2.293	2.191	1.644
Internal consistency reliability (Cronbach alpha)	.893	.694	.748	–
Kaiser-Meyer-Olkin Measure of Sampling Adequacy	.677			
Bartlett Test of Spericity Approx. Chi Square	225.847			
df	91			
sig	.000			

In sum, the 4 factors extracted from the Clients and Consultants group can be described as follows:

   i)   Factor 1: Partners' Performance
   ii) Factor 2: Partners' Permeability
   iii) Factor 3: System-based Trust
   iv) Factor 4: Relational Bonding between Partners

Contractors Group

For the Contractors Group (refer to Table 6.5), both KMO Measure of Sampling Adequacy and the Bartlett Test of Spericity indicated that the sample data from the Contractors Group was adequate for factor analysis (Sharma 1996, Hair *et al.* 1998, Cheung 1999). Applying the Eigenvalue greater than 1 rule (Sharma 1996), four trust factors were identified in the Contractors Group after VARIMAX rotation. The extracted factors accounted for 74.85% of the total variance, which was considered sufficient to explain trust with the extracted attributes.

Seven trust attributes including Unity, Problem Solving, Competent, Openness, Alignment, Information Flow and Respect were extracted for Factor 1. According to the attribute descriptions in Table 6.1, Unity, Problem Solving, Competent and Alignment are reflecting the working performance of partners. Openness and Information Flow are describing the permeability of partners. Thus, a composite factor explaining Partners' Performance and Permeability is interpreted for Factor 1 of the Contractors Group. Factor 2 extracted three attributes including Satisfactory Terms, Reputation and Adopt ADR. As explained in the results from the Clients & Consultants group, these attributes represent System-based Trust. Factor 3 consists of three attributes; Compatibility, Long-term Relations and Communication. Indeed, maintaining long-term relations and compatibility of partners should be facilitated by frequent and effective communication. Thus, the three extracted attributes represent the Relational Bonding between Partners. Factor 4 explained 11.74% of variance. A single trust attribute—Financial Stability is extracted.

To summarize, the 4 factors extracted from the Contractors Group can be described as follows:

    i) Factor 1: Partners' Performance and Permeability
    ii) Factor 2: System-based Trust
    iii) Factor 3: Relational Bonding between Partners
    iv) Factor 4: Partners' Financial Stability

Relative Importance Ranking of the Identified Factors by Factor Scores

Examining the trust factors as identified in the above PCFAs enables the understanding of trust in construction partnering in a more amenable and logical manner. It is mindful that companies are having different backgrounds and business objectives, the effectiveness of these factors may vary. In this respect, ranking the relative importance of the trust factors will further help the project team to understand, develop and enhance partners' trust more effectively. Therefore, the next step of this study was to rank the identified trust factors by their relative importance in the views of the Clients and Consultants Group and the Contractors Group. The identified trust factors were ranked by their factor scores computed by the following formula:

$$F_i = \frac{\sum_{j=1}^{n} A_{ij}}{n}$$

Where

F_i is the Factor Score

A_{ij} is the Mean Score of the jth Attribute of Factor i

The factor score of each factor was the average of the mean scores of its attributes. For example, Partners' Performance (Factor 1) of the Clients and Consultants (CLCS) Group consists of five attributes (Problem Solving, Competent, Unity, Communication and Respect). Hence, the Factor Score was computed as follow:

$$F_1(CLCS) = (4.3913 + 4.4348 + 4.3913 + 4.5652 + 4.3043)/5 = 4.4174$$

The Factor Scores for Clients & Consultants group and Contractors group were then ranked and arranged in descending order as shown in Table 6.6.

Similar factor rankings were obtained for both the Clients & Consultants and Contractors groups. System-based trust was ranked as the most important factor of trust and relational bonding ranked last. The factors lying in the middle range of importance were those related to the permeability and the performance of partners.

Table 6.6   Factor Score Ratings for Clients & Consultants Group and
             Contractors Group

Factor	Description	Factor Score	Ranking
Clients & Consultants Group			
1	Performance of Partners	4.4173	3
2	Permeability of Partners	4.7012	2
3	System-based Trust	5.1305	1
4	Relational Bonding between Partners	2.8064	4
Contractors Group			
1	Performance and Permeability of Partners	4.9133	2
2	System-based Trust	4.9404	1
3	Relational Bonding between Partners	4.1905	4
4	Financial Stability of Partners	4.6071	3

Discussion

The results from the PCFA indicate that the trust factors in construction partnering in Hong Kong is consistent with the Hartman's trust model. Trust attributes grouped in trust factors "Partners' Performance" and "Partners' Permeability" correspond to the definition of the Competent Trust and the Integrity Trust as described in the Hartman's trust model respectively. Trust attributes grouped in the factor 'Relational Bonding between Partners' also mirror Emotional Trust of the Hartman's trust model. However, as indicated from both results of the Clients & Consultants group and Contractors group, a factor called "System-based Trust" tops the ranking. The results indicate that System-based Trust is the critical component to enhance construction partner's trust in Hong Kong. According to Bulter (1983), System-based Trust refers to legally binding agreements and terms whereby trust is relied on the formalized system rather than in personal matters. This type of trust relies on social agents in society (for instance law or contract) to make partner believe their benefits can be secured in the absence of the direct personal experience (Hardin 1991, Mayer *et al.* 1995). This finding appears to be a departure to the findings of Gill and Bulter (1996) who argued that reliance on systems is a sign of mistrust. Moreover, these research results can be reconciled by examining

System-based Trust in the context of a trust development cycle. It is up to the Clients and Consultants group to decide whether to trust or mistrust the Contractors group. Installing equitable contract to avoid confrontational approach upfront can be seen as the positive gesture. Likewise, satisfaction on the contract terms would encourage the contractor to adopt a more cooperative approach. This is of particular importance during the commencement stage of the project when all partners are exposed to uncertainties. Otherwise, a mistrust cycle is more likely than a trusting one.

In addition to System-based Trust, the Contractors group also put great emphasis on Performance & Permeability of their counter parts. This is supported by the close factor scores found between the two highest ranking trust factors for the Contractors group as shown in Table 6.6 (System-based trust factor score is 4.9404 and the Performance & Permeability trust factor scores is 4.9133). These suggest that, to enhance contractor's trust, satisfactory performance and permeability of the client's team is of similar importance to system-based trust.

Similar to the results found in the Relative Importance Ranking of trust attributes, Relational Bonding between Partners with the attributes "Long-term Relations" and "Compatibility" ranked the lowest in both the Clients & Consultants Group and Contractors Group. It appears that building long-term relationship is not considered as important as it has been suggested in other reported studies (Sarker *et al.* 1998, AGCA 2002). In fact, most partnering projects in Hong Kong are for residential and commercial developments. Repetition of business between clients and contractors for such projects is not so frequent when compared with the transport military projects studied by Sarker *et al.* (1998) and AGCA (2002).

Interviews with Experts

A number of findings associated with trust in construction partnering were identified from the statistical analyzes. In order to affirm the results, confirmatory studies by interviewing 4 industry experts in construction partnering were conducted. All of them are at director level and have substantial experiences in partnering projects working in developers, consultants and contractors firms respectively. The following reports the issues identified in these interviews.

For the trust attributes, all interviewees affirmed that all of these have effect on the partners' trust level. In addition, some of the interviewees opined that "Long-term Relations" and "Compatibility" would only be applicable to partnering projects with large developers. It is because only this type of company would have a series of construction projects to facilitate longer term cooperation among partners and thus establish similar values and culture. As a matter of fact, there are only a few developers having such scale of operation in Hong Kong. Therefore, they felt that trust that are system-based or related to System-based trust, Partners' Performance and Permeability are relatively more important as these factors can be readily applied in all projects. The interviewees' opinions therefore are compatible with the findings of this study.

For the results of the PCFAs, all interviewees confirmed the significance of the identified factors in this study. In addition, the interviewees also concurred with the importance ranking of the trust factor as shown in Table 6.9. However, one of them (the Head of Procurement of a developers' company) was of the view that System-based Trust, Partners' Performance and Permeability are of equal importance. He quoted his experience in a railway project (Howlett 2002) to support his point of view. He agreed that trust can be developed if partners are satisfied on the pre-agreed contract terms with fair risk allocation and benefits sharing. However, improvement in communication, efficient information flow, working competence and the speed of problem solving of partners are also essential to enhance partner's trust during the project period. As such, the views of the interviewees were generally in line with the findings of this study. Moreover, trusts that are System-based and those related to Partners' Performance and Permeability are considered of more or less equal importance.

6.4 Summary

Developing trust among project partners is of fundamental importance for the success of partnering project. In the study reported in this chapter, views on how to develop trust from the two partners of a partnering dance were examined. Views from the experienced practitioners were first solicited and analyzed by Principal Component Factor Analysis. The factor scores derived from the analyzes were interpreted and further augmented by a confirmatory study with industry experts. In sum, the results from both the Clients & Consultants and the Contractors groups are compatible to the Hartman's trust model. Furthermore, this study pinpoints that System-based Trust is the most important trust factor. This can be explained by the fact that persistent and continual contractual relationship between a particular pair of client and contractor is rare in Hong Kong. As such, System-based Trust, which emphasizes reliance on formalized system like law and contracts, was ranked as the most important trust factor by both the Clients & Consultants and the Contractors in this Hong Kong-based study. This indicates how crucial for partners to formulate equitable contract terms and establishing channels to resolve difference right at the beginning of the project so as to trigger the trust cycle.

In addition, the findings also suggest that the role of the relational bonding on trust building is less significant when compared to other reported studies. This may also due to the fact that most partnering projects in Hong Kong are for residential and commercial developments where repeated dealings between developers and contractors are not that common.

In addition, contractors consider System-based trust and Partner's performance & permeability are of equal importance as far as trust building is concerned. The message from this finding is that the sensitivity of contractors towards trust level is strongly affected by both the system in place as well as the performance of its counterpart.

References

Bayliss, R. 2002. Project partnering—A case study on MTRC Corporation Ltd's Tseung Kwan O Extension. *HKIE Transactions* 9 (1): 1–6.

Basilevsky, A. 1994. *Statistical Factor Analysis and Related Methods: Theory and Application.* New York, Wiley.

Bernstein, C., R. J. Emerson, and A. Gabor. 1989. A Paper of Trust. http://hamp.hampshire.edu/~AWAKE/papers/891957223.html

Black, C., A. Akintoye, and E. Fitzgerald. 2000. An analysis of success factors and benefits of partnering in construction. *International Journal of Project Management* 18 (6): 423–434.

Bonet, I., B. S. Frey, and S. Huck. 2000. More order and less law: On contract enforcement, trust and crowding. *RWP00-009, Research Paper Series.* John F. Kennedy School of Government, Harvard University, USA.

Bulter, R. J. 1983. A transactional approach to organizing efficiency: Perspectives from markets, hierarchies and collectives. *Administration and Society* 15(3): 323–362.

Cheng, E. W. L., and H. Li. 2001. Development of a conceptual model of construction partnering. *Engineering, Construction and Architectural Management* 8(4): 292–303.

Cheung, S. O. 1999. Critical factors affecting the use of alternative dispute resolution processes in construction. *International Journal of Project Management* 17(3): 189–194.

Cheung, S. O., T. S. T. Ng, S. P. Wong, and C. H. Suen. 2003. Behavioral aspects in construction partnering. *International Journal of Project Management* 21(5): 333–343.

Construction Industry Institute (CII). 1989. *Partnering: Meeting the Challenges of the Future.* Construction Industry Institute, Texas, USA.

Construction Industry Review Committee of Hong Kong (CIRC). 2001. *Construct for Excellence.* Report of the Construction Industry Review Committee, Hong Kong. 52–85.

Dulaimi, M. F., F. Y. Y. Ling, G. Ofori, and N. de Silva. 2002. Enhancing integration and innovation in construction. *Building Research and Information* 30(4): 237–247

Fairholm, G. 1994. *Leadership and the Culture of Trust.* New York: Simon and Schuster.

Ford, D. 2001. *Trust and Knowledge Management: The Seeds of Success.* Queen's KBE Centre for Knowledge-Based Enterprises. http://www.business.queensu.ca/kbe

Frost, T., D.V. Stimpson, and M. R. C. Maughan. 1978. Some correlates of trust. *The Journal of Psychology* 99: 103–108.

Fukuyama, F. 1996. *Trust: The Social Virtues and the Creation of Prosperity.* New York: Free Press.

Gambetta, D. 1998. *Trust: Making and Breaking Cooperative Relations.* Oxford, Basil Blackwell.

Gill, J., and R. Bulter. 1996. Cycles of trust and distrust in joint-ventures. *European Management Journal* 14(1): 81–89.

Good, D. 1988. Individuals, interpersonal relations and trust. In *Trust: Making and Breaking Relationships,* ed. D. Gambetta and Basil Blackwell, 31–38. Oxford.

Hair, J. F., R. E. Anderson, R. E. Tatham, and W. C Black. 1998. *Multivariate Data Analysis.* 5th ed. 90–92 and 148–160. New Jersey: Prentice Hall..

Hardin, R. 1991. Trusting persons, trusting institutions. In *Strategy and Choice,* ed. R. J. Zeckhauster, 185–207. Massachusetts: MIT Press.

Hartman, F. 2003. *Ten Commandments of Better Contracting: A Practical Guide to Adding Value to an Enterprise through More Effective SMART Contracting.* ASCE Press, USA. 235–260.

Hawke, M. 1994. Mythology and reality—The perpetuation of mistrust in the building industry. *Construction Papers of the Chartered Institute of Building* 41: 3–6.

Howlett, A. M. 2002. International construction developments? What comes after partnering? *Jones Day Commentaries.* May 2002. http://www.jonesday.com/practices

Jones, G., and J. George. 1998. The experience and evolution of trust: Implications for cooperation and teamwork. *Academy of Management Review* 23(3): 531–548.

Kadefors, A. 2003. Trust in project relationships—Inside the black box. *International Journal of Project Management* 22(3): 175–182.

Kumar, N. 1996. The power of trust in manufacturer—Retailer relationships. *Harvard Business Review* (Nov–Dec Issue): 105.

Kramer, R. M., and T. R. Tyler. 1996. *Trust in Organizations—Frontiers of Theory and Research.* SAGE Publications, USA. 166–195.

Mayer, R., J. Davis, and F. Schoorman. 1995. An integrative model of organizational trust. *The Academy of Management Review* 20(3): 709–734.

Morgan, R. M., and S. D. Hunt. 1994. The Commitment-trust theory of relationship marketing. *Journal of Marketing* 58: 20–38.

Nyhan, R. C., and H. A. Marlowe, Jr. 1997. Development and psychometric properties of the organizational trust inventory. *Evaluation Review* 21(5): 614–635.

Regulatory Affairs Professional Society (RAPS). 2000. Building trust across cultural boundaries. *Regulatory Affairs Focus* (May Issue): 6–10. USA: RAPS Publication.

Romahn, E., and Hartman F. 1999. Trust: A new tool for project managers. In *Proceedings of the 30th Annual Project Management Institute 1999 Seminars and Symposium.* Philadelphia, Pennsylvania, USA.

Rosenfeld, T., A. Warszawski, and A. Laufer. 1991. Quality circles in temporary organizations, lessons from construction projects. *International Journal of Project Management* 9(1): 21–28.

Rousseau, D., S. Sitkin, R. Burt, and C. Camerer. 1998. Introduction to special topic forum. Not so different after all: A cross-discipline view of trust. *Academy of Management Review* 23(3): 393–404.

Sarker, M. B., P. S. Aulakh, and S. T. Cavusgil, 1998. The strategic role of relational bonding in inter-organizational collaborations: An empirical study of the global construction industry. *Journal of International Management* 4 (2): 415–421.

Sharma, S. 1996. *Applied Multivariate Techniques.* John Wiley & Sons, USA. 116–123.

The Associated General Contractors of America (AGCA). 2002. *Partnering Best Practice: Case Studies.* The Associated General Contractors of America, USA. http://www.partneringbestpractice.org

University of Salford and the Centre for Construction Innovation (USCCI). 2002. *Trust in Construction Review: Achieving Cultural Change.* University of Salford and the Centre for Construction Innovation, UK. http://www.scpm.salford.ac.uk/trust

William, S., G. Wood, P. McDermott, and R. Cooper. 2002. Trust in construction: Conceptions of trust in project relationships. *W92 2002 Conference–Trinidad and Tobago.* http://www.scpm.salford.ac.uk/trust

Wood, G. and P. McDermott. 1999. Searching for trust in the UK construction industry: An interim view." In *CIB W92 Procurement Systems Conference.* http://www.scpm. salford.ac.uk/trust

Wood, G., P. McDermott, and W. Swan. 2001. The ethical benefits of trust-based partnering. In *Proceedings of the Example of the Construction Industry—Business Ethics European Review Conference, UK.* http://www.scpm.salford.ac.uk/trust

Zaghloul, R, and F. Hartman. 2000. Construction contracts: The cost of mistrust. *International Journal of Project Management* 21(6): 419–424.

Trust Initiator: A Prisoner's Dilemma Perspective

Sai On Cheung
Peter Shek Pui Wong

Acknowledgements

The SEM model described in this chapter has been published in Volume 21 (2) of the *Journal of Management in Engineering*, ASCE. The study to identify trust initiator has been published in Volume 131 (10) of the *Journal of Construction Engineering and Management*, ASCE. The authors thank the ASCE for their permission to use the content therein in this chapter.

7.1 Trust in Construction Partnering

Trust is regarded not only as glue that holds contracting parties together but also the lubricant that helps to complete the project smoothly (Nicholas 1993, Whitney 1996, Wong *et al.* 2000). The happenings on a project would reinforce or reduce the trust among the people involved (Hawke 1994). Perception of trust is resulted from positive feedback towards prediction and expectation of future events. Its generation is associated with the honoring of promises by the other party (Rotter 1967). In fact, the concepts of trust are expanded in a variety of contexts (Ford 2001). For example, Rosenfeld (1991) suggested that the changing project structure and conditions, project complexity and the period of collaboration render trust relations in the construction industry differs from other settings. Thus, trust in construction partnering should be considered according to its own characteristics.

What are the factors that have an effect on trust of construction partners? From the developers' perspective, the typical determinants of the trustworthiness of contractors include: the achievement of project targets and problem-solving speed. Likewise, the promptness to honor payment certificates and the attitude on claim negotiations of an employer are often used by a contractor to gauge the trustworthiness of that employer (McGeorge and Palmer 2000). In Chapter six, fourteen trust attributes that would affect the trust level of construction partners have been identified and are listed here-follows for ease of reference:

(1) Competence of work (Competent)
(2) Problem solving ability (Problem Solving)
(3) Frequency and effectiveness of communication (Communication)
(4) Openness and integrity of communication (Openness)
(5) Alignment of effort and rewards (Alignment)
(6) Effective and sufficient information flow (Information Flow)
(7) Sense of Unity (Unity)
(8) Respect and appreciation of the system (Respect)
(9) Compatibility (Compatibility)
(10) Long-term relationships (Long-term Relations)

(11)   Financial stability (Financial)

(12)   Reputation (Reputation)

(13)   Adoption of ADR techniques (Adopt ADR)

(14)   Satisfaction of contract terms and agreements (Satisfactory Terms)

The building of trust is to enhance the achievement of project goals such as, delivery of project on time, within budget, of excellent quality, with open and honest communication, and prompt decision-making and resolution of problems (Swan *et al.* 2002). Therefore, any initiative in trust building has to respect the technical, social and commercial realities of the construction industry (Matthews 1999). Furthermore, it is imperative that project team members are willing to commit to the trust-building process.

Chapters five and six have already dealt with the fundamentals of trust, namely its functions and underpinnings and the important trust factors. As such, an overview of how trust can be achieved in co-operative contracting, and its effects on the project and its members have been provided. In this chapter, the relationship between trust and project success is first examined. An enquiry to the party most suitable to initiate trust is also explored. In both cases empirical support are provided. Again, as partnering projects exemplify co-operative contracting in construction, the study, reported in the chapter draws on data from partnering projects conducted in Hong Kong. The approaches taken to collect data can be found in Chapter six.

7.2 Structural Equation Model on Trust and Partnering Success

The Construction Industry Institute Australia (CIIA) (1996) and Larson and Drexler (1997) suggested that developing trust relations, fair benefits sharing formula, effective communication and competent management team are critical to the success of project partnering. Li *et al.* (2000) and Cheng and Li (2001) identified that top management support, mutual trust, open communication and effective coordination are the critical success factors for partnered projects. Concluded from a review on partnering studies,

Black *et al.* (2000) pinpointed that the establishment of trust among partners is the most critical factor for partnering success. As such, the above studies supported that there is a strong relationship between trust and partnering success in construction. Nevertheless, as many authors have observed, little research has explicitly related the core concept of trust specific to construction projects (Bresnen and Marshall 2000, Kadefors 2004). To examine the relationship between trust and partnering success, data collected from the questionnaire survey as reported in Chapter six were used.

Trust can be regarded as glue that fosters cooperation among organizations and an essential lubricant that helps to complete the project smoothly (Wong *et al.* 2000). Such an outcome can be obtained through seamless interaction among project partners in addressing problems and disclosing information (Kadefors 2004). The notion of trust has been expounded according to the construct and discipline characteristics (Rousseau *et al.* 1998). For example, trust in strategic alliances can be categorized as Fragile or Resilient trust. This classification focused on the relationships between the partners. Child and Faulker (1998) employed another framework and put trust in three generic classes; knowledge-based, affection-based and calculation-based thus emphasizing the source of trust. Similarly, Hartman (2003) modeled trust among project partners in three types: Competence Trust, Integrity Trust and Intuitive Trust.

In addition, Wong and Cheung (2004), in a study on ways to build trust in partnering projects, suggested a 4-factor framework to categorize trust in construction partnering projects. The four factors are Partner's Performance, Partner's Permeability, Relational Bonding and System-Based. Table 7.1 provides a comparison of the above-mentioned perspectives about trust under the 4-factor framework suggested by Wong and Cheung (2004).

7.2.1 Partners' Performance

Partners' Performance refers to the trustworthiness of a construction partner as viewed by his project team and is gauged by his own performance (USCCI 2002). According to Wong and Cheung (2004), the partner's performance can basically be examined by the following five attributes: (i) Competence of work (William *et al.* 2002),

(ii) Problem solving ability (USCCI 2002), (iii) Sense of unity (USCCI 2002), and (iv) Alignment (Wood and McDermott 1999) and (v) Respect and appreciation of the system (Gill and Bulter 1996).

Table 7.1 Factors of Trust in Strategic Alliances

Factors of Trust in Strategic Alliances			
Performance	**Permeability**	**Relational**	**System-based**
Ring (1996) Fragile Trust: Predictability of outcomes of a transaction		Resilient Trust: Predictability of goodwill of others	
Child and Faulker (1998) Knowledge-based Trust: Trust on the responsible behavior of others as their competence and good records on keeping promises		Affection-based Trust: Emotional bonds due to the lose relationships and common advantages	Calculation-based Trust: security of that party will perform as predicted because of the repressive measures.
Hartman (2003) Competence Trust: Perceptions on others ability to perform the required work. Gained by observable proofs like track record, experience etc.	Integrity Trust: Perceptions on others willingness to protect the interest of their counter parts over the project period. Gained by observable honestly and open communication	Intuitive Trust: Perceptions on others are hardly affected by the instant performance but long term relationships. Founded upon parties' prejudices and biases.	
Kadefors (2004) Calculation-based Trust: Perception on trustee to perform an action that is beneficial to the trustor. Based on the performance and competence of work.		Relational Trust: arises between parties who repeatedly interact over time and greatly affected by the emotional and personal attachments.	Institutional-based Trust: Reliance on the role of the institution like legal system to shape the conditions necessary for trust to arise.

7.2.2 Partners' Permeability

Partners' Permeability is related to the honesty and openness of the organization. This involves being open in sharing and receiving information and dealing with others in a straight forward manner (Cook and Wall 1980, Bulter 1983, Mayer *et al.* 1995, Wong *et al.* 2000). Partners' Permeability can be examined by the following trust attributes; (i) Openness and integrity of communication (Rosenfeld *et al.* 1991), (ii) Frequency and effectiveness of communication (Wood and McDermott 1999), (iii) Effective and sufficient information flow (USCCI 2002) and (iv) Financial stability (USCCI 2002).

7.2.3 Relational Bonding

Relational Bonding is identified as a trust factor as this explains trust between organizations derived from repeated interactions over time (Kadefors 2004). Relational Bonding can be examined by the following attributes: (i) Long-term relationships (Morgan and Hunt 1994) and (ii) Compatibility (Sarker *et al.* 1998).

7.2.4 System-based Trust

System-based Trust is derived from the system regulating the dealings between the contracting parties (Hardin 1991). Wong and Cheung (2004) suggest that System-based trust can be represented by three attributes: (i) Satisfaction of contract terms and agreements (Bonet *et al.* 2000), (ii) Adoption of the ADR techniques (Cheung 1999) and (iii) Reputation (Gambetta 1998). Table 7.2 summarizes all the above four factors and also provides a list of keywords for each such attribute.

The impact of the above trust factors on partner's trust level are not fully understood and are still under investigation. For example, Gill and Bulter (1996) argued that relying on System-based Trust is a feature leading to mistrust instead of enhancing the partners' trust level. Morgan and Hunt (1994) emphasized only on long-term relationships. This uni-dimensional approach takes no account of the performance of the construction partner. In these contexts, investigating the reliability of these trust factors is of both research and application interests. A better understanding of the interrelationship among trust factors and trust level shall provide valuable insights for management to devise ways and means to enhance partnering spirits.

Table 7.2   Trust Attributes in the 4-factor Framework of Trust in Construction Partnering

		Trust Attributes		References
4-Factor Framework Of Trust in Construction Partnering	**Partners' Performance**	(i) Competent	Project team members will trust each other if both of their "behaviors" and "outcomes" are competent.	William *et al.* (2002)
		(ii) Problem Solving	Construction personnel saw problem solving as an important element in building trust, especially if it is solved at the early stage.	USCCI (2002)
		(iii) Unity	Trust can be built by understanding and appreciating partners' requirements and difficulties and looking to meet partners' expectation.	*Ditto*
		(iv) Alignment	Benefits received should be fair and match with the input efforts or mistrust will result.	Wood and McDermott (1999)
		(v) Respect	A respect on the mutual dependence project management system is a source of trust.	Gill and Bulter (1996)
	Partners' Permeability	(i) Openness	Failure of integrity involves lying; cheating or hiding facts happened in the project team will tarnish trust.	Rosenfield *et al.* (1991)
		(ii) Communication	Open and frequent communication and maintaining open-door policies to each other results from a willingness of the partners to create transparency in the relationship.	Sarker *et al.* (1998)
		(iii) Information Flow	Effective and sufficient information flow would reduce risk and uncertainty of the work.	USCCI (2002)
		(iv) Financial	The financial status of the company affects the decision to trust. Contractors who have a healthy financial status are trustworthy in views of the clients as their risks to make profits by finding loopholes in contract or applying unreasonable claims are lowered down.	*Ditto*
	Relational Bonding	(i) Long-term Relations	Long term relationships among partners will lead to trust.	Morgan and Hunt (1994)
		(ii) Compatibility	Project team members will trust each other when they share similar cultures and values.	Sarker *et al.* (1998)
	System-based Trust	(i) Satisfactory Terms	Equitable agreements or contracts terms can help the contracting parties to establish trust and sustain cooperation since their perceived benefits are secured.	Bonet *et al.* (2000)
		(ii) Adoption of ADR	The implementation of ADR techniques before litigation as stated in the contract would also gain trust from other parties. These contracting parties will feel that their partners are willing to seek for win/win resolution sincerely without destroying the cooperation harmony	Cheung (1999)
		(iii) Reputation	Companies with higher reputation are more trustworthy as they do not want to lose their valuable asset.	Gambetta (1998)

Source: Wong and Cheung 2004

7.2.5 Measurement of Partnering Project Success

The purpose of partnering is to improve construction projects' performance (Egan 1998, CIRC 2001). Thus, the study on partnering success should determine whether partnering projects could achieve better performance than similar projects that do not apply a partnering approach (Bennett and Jayes 1995). Nevertheless, the Partnering Sourcing Limited (PSL) established by the Department of Trade and Industry and the Confederation of British Industry in U.K. argued that partnering is a process of continuous improvements that accumulate and very often only become apparent after project completion. In addition, some of the rewards offered by adopting partnering approach are intangible and cannot readily be measured. In this regard, it has been suggested to set project goals at different stages of a project together with procedures for their monitoring and control. If the preset targets can be achieved on or before the milestones, the effect of partnering can thus be recognized (PSL 2001). Hence, evaluating the achievements of preset project goals is accepted as an objective assessment of partnering success. Typical project goals to be met are in 5 major aspects namely: Time, Completion, Cost, Quality, Communication and Management (Munns and Bjeirmi 1996, McGeorge and Palmer 2000, Ng *et al.* 2002, Wong *et al.* 2003, Diallo and Thuillier 2004). Among the five project goals, some of them would be more related to the success of the partnering process. It depends on the preferences of the partners involved. Therefore, the relative importance of these indicators for evaluating partnering success was also investigated in this study. With this information, management can direct their resources to achieve the desired performance.

7.2.6 Hypothetical Model

The schematic model of this study is shown in Figure 7.1. The arrows represent the direction of the hypothesized influence (Bollen and Long 1992, Hoyle 1995, Molenaar *et al.* 2000). For example, the partnering success level was manifested by the achievement of project goals. Hence the arrows shown originate from the partnering success level to the time, cost, quality, communication and management. Similarly, four trust factors were represented by the fourteen trust attributes listed in Table 7.2. Thus, the arrows shown originate from the trust factors to the trust attributes.

Figure 7.1   Hypothetical Model on Trust and Partnering Success

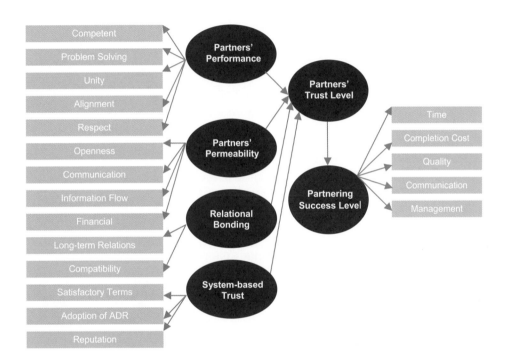

7.2.7 Data Analysis Tools

After deciding the hypothetical model of this study, the next step is to select a suitable analytical tool for testing the hypothesis. In this study, several statistical analyzes are considered. To examine the grouping of trust attributes into trust factors, Confirmatory Factor Analysis (CFA) can be used (Sharma 1996). To determine the impact of the trust factors on partner's trust level, Multiple Regression Analysis (MRA) is available (Coakes and Steed 1999, Norušis 1999).

However, when the inter-relationships of these results are to be considered in a holistic manner like the hypothetical model as shown in Figure 7.1, the technique of Structural Equation Modeling (SEM) was considered more appropriate. SEM is a multivariate technique used to estimate a series of interrelated dependent relationships simultaneously (Hair *et al.* 1998). By explicitly accounting for the measurement errors in the variables, the SEM framework produces a more accurate representation of the overall results. SEM also takes into account the errors in measurement when a large number of variables are involved (Molenaar *et al.* 2000). A computer package called LISREL 8.5 performing SEM was used in this study. SEM has been used in finding the role of relational bonding in collaboration between construction contracting organizations (Sarker *et al.* 1998) and design and build project selection (Molenaar and Songer 1998). In addition, several types of goodness-of –fit (GOF) indices are also available to assist in verifying the overall fitness of the SEM framework (Kelloway 1998).

7.2.8 Questionnaire Survey for Data Collection

The data collection questionnaire used for this study has been presented as Figure 6.1 in Chapter six. 14 questions (Q2.1 to Q2.14) were designed to collect the respondents' assessment on the degree of importance of the 14 trust attributes in developing and enhancing the trust level on their project partners using a seven-point likert scale. As described, these attributes were arranged within four-factor framework suggested by Wong and Cheung (2004) (refer to Table 7.2). The appropriateness of these groupings was first verified by a reliability test. 20 questions (Q3.1 to Q3.20) were designed to collect the respondents' assessment on the achievements of the five project goals (i.e., Time, Completion Cost, Quality, Communication and Management) in deciding partnering success. Against each of the questions, the respondents were asked to assign scores on a seven point scale. The total score of the particular goal will be averaged, to represent the achievement of the five project goals respectively. For ease of reference, the questions are listed in Table 7.3.

Table 7.3   Questions for Respondents' Assessment on the Achievements of the Five Project Goals in Deciding Partnering Success

Achievements of Project Goals in	References
Time (Average Score)	
The project can meet the committed target date for completion as stated in the partnering charter	Gattorna and Walter(1996), Ng *et al.* (2002)
Resolution of problems become more efficient	Naoum (2003)
The work process throughout the project is smooth and efficient and the redundant work is eliminated	Ditto
Stakeholders can make decisions efficiently in problem solving	Li *et al.* (2001)
Completion Cost (Average Score)	
The project can meet the committed target budget for completion as stated in the partnering charter	Gattorna and Walter(1996), Ng *et al.* (2002)
Cost saving proposals efficiently lead clients and contractors to achieving higher profit margins	Naoum (2003)
Effective utilization of resources reduces wastage of both labor and materials	Ditto
Reduced claims, variations and risk of litigation	Gattorna and Walter(1996), Ng *et al.* (2002)
Quality (Average Score)	
The project can meet the committed target quality as stated in the partnering charter	Black *et al.* (2000), Gattorna and Walter(1996), Ng *et al.* (2002)
Continuous improvement of quality can be achieved by joint evaluation and implementation of innovative or effective methods.	Ditto
Continuous improvement of safety levels can be achieved by joint evaluation	Naoum (2003)
Stakeholders will accept others' mistakes and consider remedying the fault together	Ng *et al.* (2002)
Communication (Average Score)	
The meetings are frequent, sufficient and effective	Black *et al.* (2000)
All stakeholders' interests are considered and respected during discussion	Ng *et al.* (2002)
Problems are recognized and rectified at the earliest stage through meetings or other communication	Wright (1997)
Sharing of information is open, honest, accurate, timely, helpful and trust manner	Ng *et al.* (2002)
Management (Average Score)	
A good leader chairs the partnering meeting, monitoring and supporting the functionality of partnering	Wright (1997)
All representatives involved are of sufficient caliber to make agreeable decisions	Ditto
Partnering tools including problem identification, conflict escalation procedures, ADR approaches and an evaluation methodology.	Naoum (2003), Ng *et al.* (2002)
Mutual goals can be achieved and mutual trust can be enhanced	Naoum (2003)

7.2.9 Structural Equation Modeling (SEM) on Trust and Partnering Success

Theoretically, a structural equation model comprises two types of models: a measurement model and a structural model. The measurement model is concerned with how well the variables measure the latent factors, thus addressing the reliability and validity of a model. The structural model is concerned with the relationships between the latent factors by describing the amount of explained and unexplained variance. It is akin to the system of simultaneous regression models (Molenaar *et al.* 2000). In this study, trust attributes were grouped to explain four latent trust factors; the trust factors were then grouped to explain the partners' trust level. In addition, the average score of the five project goals were grouped to explain the partnering success level.

7.2.10 The Initial SEM Specification

A SEM specification should firstly be built on the theoretical expectations and findings from literature review (Molenaar et al. 2000, Gainey and Klass 2003). Therefore, in this study, the initial SEM specification followed the hypothetical model as shown in Figure 7.1. In addition, to ensure the appropriateness of groupings of the identified trust attributes into trust factors, Cronbach alphas reliability testing was applied (Peter 1981, Sharma 1996). The alpha value ranges from 0 to 1. The values ranging from 0.6 to 0.7 are regarded as "sufficient" and the value higher than 0.7 is regarded as "good" in reliability testing (Sharma 1996). Referring to Table 7.4, all groupings in the Initial SEM have a Cronbach alpha values higher than 0.6. This indicates that the reliability of the variable groupings in the Initial SEM.

7.2.11 SEM Refinements

The overall fitness of the Initial SEM can be assessed by examining the Goodness of Fit (GOF) indices. In fact, there are several GOF indices available to test the fitness of the SEM. If the GOF indices of the Initial SEM could not reach the recommended levels, model refinements are required to improve the overall fitness. The recommended levels of the GOF indices are as shown in Table 7.5. In this study, model refinements were performed by two methods. Firstly, this was done by a systematic approach in eliminating low correlation paths and associated variables (Churchill 1979, Sarker *et al.* 1998). Secondly, this was done by revising the inter-relationships paths or adding covariance error paths between the variables or latent factors. This method was needed to refine the SEM model with reference to the modification indices provided by the LISREL 8.5 programme. After refinements, the model that performed well in terms of both GOF indices and the theoretical expectations was selected as the Final SEM model (Molenaar *et al.* 2000).

Table 7.4   Reliability Testing of the Initial SEM

Initial SEM Grouping	Item	Cronbach Alpha Value (α)
Trust Factor 1 (Partners' Performance)	Competent	.8864
	Problem Solving	
	Unity	
	Alignment	
	Respect	
Trust Factor 2 (Partners' Permeability)	Openness	.7128
	Communication	
	Information Flow	
	Financial	
Trust Factor 3 (Relational Bonding)	Long-term Relations	.7402
	Compatibility	
Trust Factor 4 (System-based trust)	Satisfactory Terms	.6282
	Adoption of ADR	
	Reputation	
Partnering Success Indicators	Time (Average Score)	.9346
	Completion Cost (Average Score)	
	Quality (Average Score)	
	Communication (Average Score)	
	Management (Average Score)	

7.2.12 Results in the SEMs

The Initial SEM based on Figure 7.1 was analyzed by LISREL 8.5. According to the GOF indices of the Initial SEM, model refinement was required until the indices reach the recommended levels as shown in Table 7.5. After three refinements through eliminating and relocating some trust attributes, latent trust factors and partnering success indicator as shown in the remarks of Table 7.5, the GOF indices of the fourth model achieved the recommended levels. The final model using standard the SEM terminology and graphical notation is presented in Figure 7.2.

Table 7.5   Results of GOF Measures

Goodness-of-Fit (GOF) Measure	Recommended Level of the GOF Measure	Starter SEM	Final SEM[a]
X^2/degree of freedom	Recommended Level from 1 to 2	2.00	1.87
Goodness of Fit Index (GFI)	0 (no fit) to 1 (perfect fit)	0.66	0.84
Root Mean Square Error of Approximation (RMSEA)	<0.05 indicates a very good fit, the threshold level is 0.10	0.11	0.08
p-value for testing closeness of fit (RMSEA <0.05)	p-value for hypothesis testing of RMSEA	0.00	0.04
Tucker-Lewis Index (NNFI)	0 (no fit) to 1 (perfect fit)	0.69	0.86
Normal Fit Index (NFI)	0 (no fit) to 1 (perfect fit)	0.60	0.76

Source: adopted from Molenaar et al. 2000

[a] Remarks:

i.   Deleted trust attribute: Financial, Compatibility, Long-term Relations, Reputation

ii. Deleted latent trust factor: Relational Bonding

iii. Deleted partnering success indicator: Time (Average Score)

iv. Relocated trust attribute: Alignment relocated from latent factor "Partners' Performance" to latent factor "System-Based trust"

As the SEM depicts a system of regression equations (Molenaar *et al.* 2000), there was a squared multiplied correlation (R-square) associated with the error term in each equation as shown in Figure 7.2. The error terms represent the portion of the variables that are not explained. Since the R-square value between three trust factors and the trust level is 0.72, the SEM explains about 72% of the variability in trust level.

Figure 7.2 Final SEM with Squared Multiple Correlations (R-square) and Error Terms

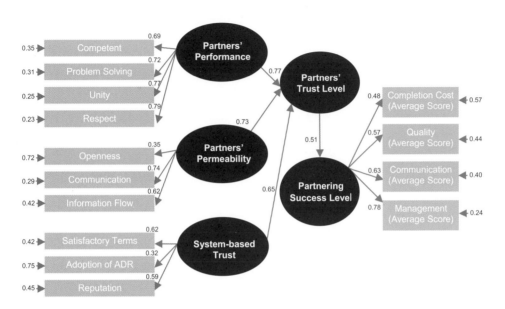

The final model was compared with the Initial SEM, 4 variables and 1 latent factor were eliminated due to their low correlations with the other latent factors in the Final SEM. As such, three factors namely: Partners' Performance, Partners' Permeability and the System-based Trust remain in the model. Relational bonding does not appear to have high correlation with the partners' trust level. The summary of standardized coefficients of the final model is shown in Table 7.6. All path coefficients are positive and significant at p <.05 and thus their significance to the model are augmented.

The GOF indices are essential tools for assessing the fitness of SEMs. From Table 7.5 the model fitness of the final SEM for trust and partnering success is supported by the results of the indices. The ratio for $\chi 2/df$ is 1.87 (186.68/100) and the Goodness of Fit index value is 0.84, both indices indicate that the final model provides a good fit to the data. The Root Mean Square Error of Approximation (RMSEA) value is 0.08 at p<0.05, this indicates the final model cannot be rejected at a high level of confidence. In addition, both Normal Fit Index (NFI) and the Tucker-Lewis Index (NNFI) values support the statistical fit between the measurement model and the data (Gainey and Klass 2003, Molenaar *et al.* 2000; Sarker *et al.* 1998).

Table 7.6   Standardized Coefficient Values of Paths in the Final SEM

Path	p-value*
Partners' Performance → Competent	0.87
Partners' Performance → Problem Solving	0.89
Partners' Performance → Unity	0.92
Partners' Performance → Respect	0.93
Partners' Permeability → Openness	0.62
Partners' Permeability → Communication	0.90
Partners' Permeability → Information Flow	0.86
System-based Trust → Satisfactory Terms	0.82
System-based Trust → Adoption of ADR	0.59
System-based Trust → Alignment	0.80
Partners' Performance → Partners' Trust Level	0.88
Partners' Permeability → Partners' Trust Level	0.85
System-based Trust → Partner's Trust Level	0.81
Partner's Trust Level → Partnering Success Level	0.72
Partnering Success Level → Completion Cost (Ave. Score)	0.73
Partnering Success Level → Quality (Ave. Score)	0.76
Partnering Success Level → Communication (Ave. Score)	0.80
Partnering Success Level → Management (Ave. Score)	0.93

*All standardized coefficient values are significant at $p < 0.05$

7.2.13 Discussion

The Final SEM suggested that three trust factors have significant correlation on Partners' Trust Level: Partners' Performance, Partners' Permeability and System-based Trust. Relational bonding, among the four trust factors, was identified as the least important factor. This marked a departure from the generally accepted view that partnering is built on relations. This may have been due to the fact that most partnering projects in Hong Kong are obtained through competitive bids. Further collaboration between developers and contractors in other projects could not be guaranteed if compared with other partnering studies conducted with projects secured through negotiated bids (Sarker *et al.* 1998, AGCA 2002).

A significant and positive relationship (standardized coefficient = 0.88) was found between the Partners' Performance and the Partners' Trust Level. This suggested that partner's trust is mainly enhanced by the performance of his counter part. According to the measurement model of the Final SEM, Partners' Performance was measured by four trust attributes: Competence, Problem Solving Ability, Unity and Respect. In fact, evaluating performance is a means for partners to judge the trustworthiness of others. This is because performance can more readily and tangibly be evaluated by the contracting parties (Kadefors 2003).

Partners' Permeability was also found to be a critical trust factor (standardized coefficient = 0.85). It was directly reflected by three trust attributes namely: Communication, Openness and Information Flow. These attributes led to greater trust because as the better the partners interact and exchange quality information, the more they understand each other's needs thereby behavior norms between them can be established (McAllister 1990, Gainey and Klass 2003).

The SEM also suggested that system-based trust is an essential trust factor of partner (standardized coefficient = 0.81). System-based trust has three attributes namely Adoption of ADR, Alignment and Satisfactory Terms. In fact, this finding suggested that the clarity and the fairness of the contract terms would help partners to kick start the trust cycle, especially at the beginning of the project when both developers and contractors are exposed to uncertainties. On the other hand, if rewards and fairness cannot be secured by the system-based trust, the mistrust cycle begins (Bonet *et al.* 2000).

Strong correlations were also found between the Partnering Success Level and the achievement of project goals on Completion Cost (standardized coefficient = 0.76), Quality (standardized coefficient = 0.73), Communication (standardized coefficient = 0.80) and Management (standardized coefficient = 0.93). However, Time is eliminated in the final SEM. This finding also does not sit comfortably with the commonly accepted view that meeting the goals on time is critical for partnering success (McGeorge and Palmer 2000). Nevertheless, the study was conducted in Hong Kong around a period of economic recession (from 1998 to mid-2003). The decreasing gross margins of the real estate development tempered the incentive of the developers to complete the development projects in a faster pace (C & S Dept., HKSAR 2004). Hence, the above phenomenon may help to explain why meeting time target is not a prime measure of partnering success in Hong Kong at the time of the study.

The values of the standardized coefficients suggest that Management and Communication are important goals to be achieved for the purpose of facilitating partnering success. This is in line with the findings that the primary function of adopting partnering is to provide a platform for contracting parties for active communication and effective management in order to reduce conflicts and prevent escalation of the problems (Li *et al.* 2001). The strong relationship between Partners' Trust Level and the Partnering Success Level is augmented by the high standardized coefficient value of 0.72 between them. Final SEM supports the hypothesis that the success of partnering projects in Hong Kong is significantly correlated to the trust levels among the project partners.

7.3 Trust Initiator: A Prisoner's Dilemma Perspective

Skeptics about the use of partnering in construction often raise two questions: "Are co-operative moves possible in a conflicts environment like that of construction?" and "Who is willing to instigate the first co-operative move?" These questions can be framed as the prisoner's dilemma (PD) game suggested by Rapoport and Chammah (1965) (Deutsch 1960, La Porta *et al.* 1997, James 2002). They described PD as a two-party non-constant-sum game in which both parties prefer some outcomes to the other. Moreover, the outcomes depend on the reaction of the other party. In the context of construction partnering, the moves of partners can either be competitive or co-operative. Competitive moves are those that focus on partner's own interest. This generally would invoke retaliation and/or defensive responses. Co-operative moves are those that put the interest of both partners first. Co-operative moves are characterized by reciprocal moves. This means if a partner is behaving co-operatively, he is expecting a co-operative response from his counter part. This expectancy is built on trust.

Unfortunately, the desire of a partner to select a co-operative move is usually low (James 2002). As illustrated in Figure 7.3, Partner A may either trust or mistrust Partner B, who in turn can either honor or exploit the trust. When both partners co-operate by selecting the (trust, honor) pair, they both can be benefited. If Partner A provides a co-operative move but Partner B exploits it (trust, exploit), then Partner A loses their cost invested on trust (i.e., $-c_a$). On the other hand, Partner B receives a reward of $(r + p)$ where p represents a premium that he earned from not putting effort on honoring trust expected by Partner A. If Partner A mistrusts and Partner B honors it, then Partner B incurs a cost of $-c_b$. If both parties select competitive moves (mistrust, exploit), their investment costs on trust are both 0. From Figure 7.3, it is can be seen that if Partner B is to maximize his rewards, he will exploit the trusting behavior of Partner A. Hence, the incentive for Partner A to trust Partner B is low if he anticipates that Partner B adopts a dominant strategy to exploit trust. In this context, the famous Nash equilibrium of (mistrust, exploit) will be resulted (James 2002). This scenario is commonly perceived in construction where nobody is considered to be trustworthy. This framework thus helps to explain why trust is so difficult to be built in construction projects (Hawke 1994).

Figure 7.3   Theoretical Model of Developing Trust in Construction Partnering

		Partner B	
		Honor	Exploit
Partner A	Trust	r, r	$-c_a$, r+p
	Mistrust	0, $-c_b$	0, 0

Source: adopted from James 2002

With trust being the most critical success factor for partnering (CII 1989, Black *et al.* 2000, CIRC 2001, Bayliss 2002), it is not difficult to understand why despite the potential benefits that can be derived, partnering fails to deliver if partners have low confidence to initiate trust. In this connection, James (2002) advocates that creating appropriate incentives for co-operative play could alter the orientation of partners from mistrust to trust. He suggests four ways to create trust incentives. These include writing an explicit contract, relying on reward and punishment, repeating the interaction and emphasizing on the honesty among parties. These are good strategies, the next question is: "who is best to be the first trust mover? [i.e., the partner who kick-starts the trust cycle?]" In addition, Kenneth and Martin (2001) suggest that if the first mover selects to trust, a prospective co-operation cycle begins and thus facilitates the partnering success. Their point of view was also echoed by Gill and Bulter (1996) who found that trust cycle begins when one of the partners genuinely trust the other. In sum, the prisoners' dilemma framework illustrates that trust is built on reciprocal trust moves. However, the construction industry is dominated by mistrust orientations. To overcome this inertia, the trust driver needs first to be identified.

This section of the chapter presents a study that seeks to enhance the understanding on construction partnering by identifying candidate for trust driver, client or contractor, in construction partnering. Data for this study were collected from the questionnaire survey as reported in Chapter six.

7.3.1 Explore the Candidate for Trust Driver

To explore which partner would best be the trust driver, multiple regression analysis was used. Multiple regression analysis is typically used to find the best prediction of a dependent variable (trust level) from several independent variables (trust factors) by an equation. The general equation of multiple regression (Norušis 1999, Coakes and Steed 1999) is shown as follows:

$$Y = a + b_1X_1 + b_2X_2 + b_3X_3 \ldots \quad + b_nX_n + \varepsilon \quad \text{Where:}$$

Y	= Dependent variable (trust level)
$X_1, X_2 \ldots X_n$	= Independent variables (trust factors)
$a, b_1, b_2 \ldots b_n$	= Unknown constant
ε	= Random error for any given set of values for $X_1, X_2 \ldots X_n$

There are three major regression models; they are standard or simultaneous regression, hierarchical regression and stepwise regression model. This research study adopted standard regression because this method considers the contributions of all predictors in explaining the dependent variables (Norušis 1999, Coakes and Steed 1999). The R-square value in the regression analysis represents the power of the independent variables to explain the dependent variable. The value can vary from 0 to 1. If the value is 1.0, it is indicated that all the change in the dependent variable would be due to the changes of the independent variables and suggests a close relationship between the dependent and the independent variables (Hair *et al.* 1998). This relationship can be used to suggest a candidate for trust driver. If the relationship is high, this can be interpreted that the trusting moves of the partner have a significant correlation with the trust level of the partner as perceived by the responding group, who will be more inclined to provide reciprocal trusting moves. On the other hand, if the correlation is not that significant, this would suggest the trusting moves of the partner are not that effective in deriving trust on his counterpart. As such, his suitability to assume the role of trust drive is less apparent.

7.3.2 Discussion

Critical Trust Factors

As reported in Chapter six, trust factors of the two groups of participants in the partnering projects: a) Client/Consultant and b) Contractor have been identified through the Principal Component Factor Analyzes. They are listed in Table 7.7 for ease of reference:

Table 7.7   Trust Factors of the Two Groups of Participants in the Partnering Projects

	Client/Consultant	Contractor
Factor 1	Partners' performance	Partners' performance and permeability
Factor 2	Partners' permeability	System-based trust
Factor 3	System-based trust	Relational bonding between partners
Factor 4	Relational bonding between partners	Partners' financial stability

Trust Driver: Client/Consultant or Contractor

Having identified the trust factors, the next step is to investigate which partner is more appropriate to act as the trust initiator. For this purpose, the impact of the trust factors on trust level shall be instrumental. As discussed, comparing the R-square values of the multiple regression equations of partners would assist in suggesting the trust driver in construction partnering. In this part of the study, trust factor scores were set as the independent variables, while the partner's trust level was set as a dependent variable in the Multiple Regression Analyzes. The trust factor score is an average score of its respective attributes. Take Factor 1 of the Client/Consultant group as an example, the scores of the five trust attributes: Problem solving, Competent, Unity, Communication and Respect were averaged as follows:

$$F_1(\text{Client/consultant}) = (4.3913 + 4.4348 + 4.3913 + 4.5652 + 4.3043) / 5 = 4.4174$$

Client/Consultant Group

The statistical results of the Multiple Regression Analysis of the Client/Consultant group were shown in Table 7.8.

Table 7.8   Multiple Regression Model (Partner's Trust Level vs Extracted Trust Factor)— Client/Consultant Group

Independent Variables (trust factors)	Un-standardized Coefficients		Standardized Coefficients			
	B	Std. Error	Beta	t-value	Sig.	R^2
(Constant)	4.524	.075		60.208	.000	.930
Partners' performance	.975	.077	.835	12.659	.000	
Partners' permeability	.542	.077	.464	7.038	.000	
System-based trust	.130	.077	.111	1.685	.111	
Relational bonding between partners	-8.203E-02	.077	-.070	-1.065	.303	

Dependent Variable: Trust Level on Contractor

The R-square value is 0.930, meaning that the Multiple Regression equation can explain 93% of the total variance. This indicates strongly the effectiveness of the Client/Consultant's trust factors to motivate their trust level (Hair *et al.* 1998). The standardized regression coefficient (beta value) of "Partners' performance" (Factor 1) is 0.835, "Partners' permeability" (Factor 2) is 0.464, "System-based trust" (Factor 3) is 0.111 and "Relational bonding between partners" (Factor 4) is -0.070. The standardized regression coefficient (beta value) denotes the estimated change in independent variable (the trust factor) for a unit change in the dependent variable (trust level). From the results, it is observed that "Partners' performance" (Factor 1) and "Partners' permeability" (Factor 2) have a higher beta value with the significance level at $p \leq 0.001$. Thus, when the contractor is providing trusting moves by performing competently and maintaining an open and effective communication system, the corresponding increase in trust on them by the Client/Consultant is apparent.

Contractors Group

The statistical results of the contractors group are shown in Table 7.9.

Table 7.9 Multiple Regression Model (Partner's Trust Level vs Extracted Trust Factor)— Contractor Group

Independent Variables (trust factors)	Un-standardized Coefficients		Standardized Coefficients			
	B	Std. Error	Beta	t-value	Sig.	R^2
(Constant)	5.500	.142		38.780	.000	.482
Partners' performance	.565	.144	.587	3.911	.001	
Partners' permeability	9.725E-02	.144	.101	.673	.507	
System-based trust	-.170	.144	-.177	-1.117	.251	
Relational bonding between partners	.298	.144	.310	2.062	.051	

Dependent Variable: Trust Level on Client/Consultant

The R-square value is 0.482, meaning that the Multiple Regression equation can explain 48.2% of the total variance. Compared with the results in the Client/Consultant group, the effectiveness of Client/Consultant trusting moves in fostering trust on them by the contractors is much lower. The standardized regression coefficient (beta value) of "Partners' performance and permeability" (Factor 1) is 0.587, "System-based Trust" (Factor 2) is 0.101, "Relational bonding between partners" (Factor 3) is -0.177 and

"Partners' financial stability" (Factor 4) is 0.310. As indicated, only "Partners' performance and permeability" (Factor 1) is effective to motivate contractors' trust level at significance level p ≤ 0.001. As discussed, the significant predictive power of trust factors on trust score as indicated in Table 7.9 suggests the relative effectiveness of the contractors to act as trust driver in construction partnering.

7.4 Summary

Structural Equation Modeling (SEM) technique was employed to examine and confirm the positive relationship between trust level and partnering success. The results enable management to gain better understanding and valuable insights to devise ways and means to re-focus their partnering objectives and thus ensure success. The final SEM supports the hypothesized positive relationships between Partners' trust level and partnering success. Partners' Performance, Partners' Permeability and System-based Trust were found to be the factors that were strongly linked to Partners' Trust Level. Final SEM placed lower importance of relational bonding towards Partners' Trust Level. This finding negates the generally accepted view that partnering is built on relations. This may due to the fact that most partnering projects in Hong Kong were obtained through competitive bids. Chances for further collaboration between developers and contractors for such projects were not so great if compared with other partnering studies conducted with projects secured through negotiated bids (Sarker *et al.* 1998, AGCA 2002). In addition, achieving time target is found to be less important in facilitating partnering success. This also marks a departure from the conventional belief that achieving the time target is a major success indicator. It is opined that this finding is attributed to the downstream of the Hong Kong economy which lowered the incentive of both public and private developers to complete their developments earlier.

The SEM model is built on the belief that trust is the fundamental success factor for construction partnering. By comparing the standardized regression coefficients (beta values) of the regression equations, the critical trust factors for both groups are basically the same: Performance and Permeability. Hence, to cultivate trust among the contracting partners, it is necessary to perform competently and communicate openly and effectively. Performance is evaluated by the problem solving ability and competence of work. Permeability is often assessed by the effectiveness and efficiency of the communication between the construction partners. The importance of performance is self-explanatory as clinical performance underpins project success. Without competent performance, trust can never be established. Communication is essential to resolve difference efficiently and expeditiously. In fact, an effective communication system can avoid problems becoming disputes.

The Prisoner's Dilemma framework explains that trust cycle needs to be kick started by a trust driver, a situation conceived by many as impossible in the construction industry. Trust factors have been extracted from a principal component factor analysis reported in Chapter six. The technique of multiple regression is employed with these trust factors as independent variables and trust level as the dependent variable to consider a candidate for the trust initiator. This was applied to the two groups of data: Client/Consultant and Contractor. The R-square value in the regression equation represents the power of the independent variables to explain the dependent variable. The value can vary from 0 to 1. If the value is 1.0, it is indicated that all the change in the dependent variable would be due to the changes of the independent variables and suggests a close relationship between the dependent and the independent variables (Hair *et al.* 1998).

It is found that trusting moves of Contractors have a significant correlation with the trust level of them as perceived by the Client/Consultant. This suggests that if trusting moves are initiated by a contractor, there are good chances that reciprocal trusting moves from the client will be returned. In this respect, a trust cycle can be spun off. The relative lower predictive power of Clients'/Consultant trusting moves towards trust on them (by the Contractor) suggests that contractors are more cautious towards trusting moves from the client. In these perspectives, the findings suggests that in construction partnering, if a contractor perform and maintain an effective communication system, trust from the client and consultant can be expected. With this, a trust cycle is ready to spin off upon reciprocal trusting moves from the client/consultants. As such, contractor is a suitable candidate as the trust driver in a partnering endeavor.

References

Bayliss, R. 2002. Project partnering—A case study on MTRC Corporation Ltd's Tseung Kwan O Extension. *HKIE Transactions* 9(1): 1–6.

Bennet, J., and S. Jayes. 1995. *Trusting the Team—The Best Practice Guide to Partnering in Construction.* University of Reading. Centre for Strategic Studies in Construction. Reading, U.K.

Black, C., A. Akintoye, and E. Fitzgerald. 2000. An analysis of success factors and benefits of partnering in construction. *International Journal of Project Management* 18(6): 423–434.

Bollen, K. A., and J. S. Long. 1992. Tests for structural equation models: Introduction. *Social Methods Research* 21: 123–131.

Bonet, I., B. S. Frey, and S. Huck. 2000. More order and less law: On contract enforcement, trust and crowding. In *RWP00-009, Research Paper Series.* John F. Kennedy School of Government, Harvard University, USA.

Bresnen, M., N. Marshall. 2000. Partnering in construction: A critical review of issues, problems and dilemmas. *Construction Management and Economics* 18(2): 819–832.

Bulter, R. J. 1983. A transactional approach to organizing efficiency: Perspectives from markets, hierarchies and collectives. *Administration and Society* 15(3): 323–362.

Census and Statistics Department. 2004. *Gross Domestic Product (GDP) by Economic Activity at Current Prices.* Census and Statistics Department, HKSAR. http://www.info.gov.hk/censtatd/eng/hkstat/fas/nat_account/gdp

Cheng, E. W. L., and H. Li. 2001. Development of a conceptual model of construction partnering. *Engineering, Construction and Architectural Management* 8(4): 292–303.

Cheung, S.O. 1999. Critical factors affecting the use of alternative dispute resolution processes in construction. *International Journal of Project Management* 17(3): 189–194.

Cheung, S. O., T. S. T. Ng, S. P. Wong, and C. H. Suen. 2003. Behavioral aspects in construction partnering. *International Journal of Project Management* 21(5): 333–343.

Child, J.,and D. Faulker. 1998. *Strategies and Cooperation: Managing Alliances, Network and Joint Ventures*. Oxford: Oxford University Press, U.K.

Churchill, G. A. Jr. 1979. A paradigm for developing better measures of marketing constructs. *Journal of Marketing Research* 46: 64–73.

Construction Industry Institute (CII). 1989. *Partnering: Meeting the Challenges of the Future.* Construction Industry Institute, Texas, USA.

Construction Industry Institute Australia (CIIA). 1996. *Partnering: Models for Success, Research Report No. 8.* Construction Industry Institute Australia, Australia.

Coakes S. J., and L.G. Steed. 1999. *SPSS: Analysis without Anguish: Versions 7.0, 7.5, 8.0 for Windows*. Brisbane, and Jacaranda Wiley. 167–179.

Construction Industry Review Committee of Hong Kong (CIRC). 2001. *Construct for Excellence*. Report of the Construction Industry Review Committee, Hong Kong. 52–85.

Cook, J., and T. Wall. 1980. New work attitude measures of trust, organizational commitment and personal need non-fulfillment. *Journal of Occupational Psychology* 7(1): 1–16.

Deutsch, M. 1960. Trust, trustworthiness and the F-scale. *Journal of Abnormal and Social Psychology* 61: 138–140.

Diallo, A., and D. Thuillier. 2004. The success dimensions of international development projects: The perceptions of African project coordinators. *International Journal of Project Management* 22(1): 19–31.

Egan, J. 1998. *Rethinking Construction: The Report of the Construction Task Force.* Department of the Environment. Transport and the Regions. London U.K.

Ford, D. 2001. *Trust and Knowledge Management: The Seeds of Success.* Queen's KBE Centre for Knowledge-Based Enterprises. http://www.business.queensu.ca/kbe

Gainey, T. W., and B. S. Klass. 2003. The outsourcing of training and development: factors impacting client satisfaction. *Journal of Management* 29(2): 207–229.

Gambetta, D. 1998 *Trust: Making and Breaking Cooperative Relations.* Oxford, Basil Blackwell.

Gattorna, J. L., and D.W. Walter. 1996. *Managing the Supply Chain*. New York. Macmillan. 189–203.

Gill, J., and R. Bulter. 1996. Cycles of trust and distrust in joint-ventures. *European Management Journal* 14(1): 81–89.

Hair, J. F., R. E. Anderson, R. E. Tatham, and W. C. Black. 1998. *Multivariate Data Analysis.* 5th ed. 90–92 and 148–160. Prentice Hall, New Jersey, USA.

Hardin, R. 1991. Trusting persons, trusting institutions. In *Strategy and Choice,* ed. R. J. Zeckhauster, 185–207. Massachusetts: MIT Press.

Hartman, F. 2003. *Ten Commandments of Better Contracting: A Practical Guide to Adding Value to an Enterprise through More Effective SMART Contracting.* ASCE Press, USA. 235–260.

Hawke, M. 1994. Mythology and reality—The perpetuation of mistrust in the building industry. *Chartered Institute of Building* 41: 3–6.

Hoyle, R. H. 1995. *Structural Equation Modeling: Concepts, Issues and Applications.* Ed. R. H. Hoyle. Sage, Thousand Oaks, Calif., USA.

James, S. 2002. The trust paradox: A survey of economic inquiries into the nature of trust and trustworthiness. *Journal of Economic Behavior and Organization* 41: 291–307.

Kadefors, A. 2003. Trust in project relationships—Inside the black box. *International Journal of Project Management* 22(3): 175–182.

Kelloway, E. K. 1998. *Using LISREL for Structural Equation Modeling: A Researcher's Guide.* SAGE Publications, International Education and Professional Publishers, USA. 8–30.

Kenneth, C., and S. Martin. 2001. The sequential prisoner's dilemma: Evidence on reciprocation. *Economic Journal* 111(468): 51–68.

La Porta, R., F. Lopez-de-Silanes, A. Shleifer, and R.W. Vishny (1997). Trust in large organizations. *American Economic Review Papers and Proceedings* 87(2): 333–338.

Larson, E. W., and J. A. Drexler. 1997. Barriers to project partnering: Report from the firing line." *Project Management Journal* 28(1): 46–52.

Li, H., E. W. L. Cheng, and P. E. D. Love. 2000. Partnering research in construction. engineering. *Construction and Architectural Management* 7(1): 76–92.

Li, H., E. W. L. Cheng, P. E. D. Love, and Z. Irani. 2001. Co-operative benchmarking: A tool for partnering excellence in construction. *International Journal of Project Management* 19(3): 171–179.

Matthews, J. 1999. Applying partnering in the supply chain. *Procurement Systems: A Guide to Best Practice in Construction.* E. & F. N. Spon, London. 252–275.

Mayer, R., J. Davis, and F. Schoorman. 1995. An integrative model of organizational trust. *The Academy of Management Review* 20(3): 709–734.

McAllister, D. J. 1990. Affect and cognition based trust as foundations for interpretation. *Academy of Management Journal* 38(1): 24–32.

McGeorge D., and A. Palmer. 2000. *Construction Management New Directions*. Blackwell Science. 188–239.

Molenaar, K., and A. D. Songer. 1998. Model of public sector design-build project selection. *Journal of Construction Engineering and Management* 124(6): 467–479.

Molenaar, K., S. Washington, and J. Diekmann. 2000. Structural equation model of construction contract dispute potential. *Journal of Construction Engineering and Management* 126(4): 268–277.

Morgan, R. M., and S. D. Hunt. 1994. The commitment-trust theory of relationship marketing. *Journal of Marketing* 58: 20–38.

Munns, A. K., and B. F. Bjeirmi. 1996. The role of project management in achieving poject success. *International Journal of Project Management* 14(2): 81–87.

Nicholas, T. 1993. *Secrets if Entrepreneurial Leadership—Building Top Performance through Trust and Teamwork*. Dearborn Publishing.

Ng, S. T., T. M. Rose, M. Mak, and S. Chen. 2002. Problematic issues associated with project partnering—The contractor perspective. *International Journal of Project Management* 20(6): 437–449.

Naoum, S. 2003. An overview into the concept of partnering. *International Project of Project Management* 21(1): 71–76.

Norušis, M. J. 1999. *SPSS 9.0 Guide to Data Analysis*. New Jersey, Prentice Hall, USA. 373–430.

Partnering Sourcing Limited (PSL). 2001. *A Director's Guide to Partnering*. Partnering Sourcing Limited, Department of Trade and Industry and the Confederation of British Industry, U.K. http://www.pslcbi.com/PSL_guides

Peter, J. P. 1981. Reliability: A review of psychometric basics and recent marketing practices." *Journal of Marketing Research* 16: 6–17.

Rapoport, A., and A. M. Chammah. 1965. *Prisoner's Dilemma*. Ann Arbor: University of Michigan Press.

Ring, P. S. 1996. Fragile trust and resilient trust and their roles in cooperative inter-organizational relationships. *Business & Society* 35(2): 148–175.

Rosenfeld, T., A. Warszawski, and A. Laufer. 1991. Quality circles in temporary organizations, lessons from construction projects. *International Journal of Project Management* 9(1): 21–28.

Rotter, J. B. 1967. A new scale for the measurement of inter-personal trust. *Journal of Personality* 35: 1–7.

Rousseau, D., S. Sitkin, R. Burt, and C. Camerer. 1998. Introduction to special topic forum. Not so different after all: A cross-discipline view of trust. *Academy of Management Review* 23(3): 393–404.

Sarker, M. B., P. S. Aulakh, and S. T. Cavusgil. 1998. The strategic role of relational bonding in inter-organizational collaborations: An empirical study of the global construction industry. *Journal of International Management* 4(2): 415–421.

Sharma, S. 1996. *Applied Multivariate Techniques*. John Wiley & Sons, USA. 116–123.

The Associated General Contractors of America (AGCA). 2002. *Partnering Best Practice: Case Studies*. The Associated General Contractors of America, USA. http://www.partneringbestpractice.org

Swan, W., G. Wood, P. McDermott, and R. Cooper. 2002. Trust in construction: Conceptions of trust in project relationship. In *W92 2002 Conference*. www.scpm.salford.ac.uk/trust

University of Salford and the Centre for Construction Innovation (USCCI). 2002. *Trust in Construction Review: Achieving Cultural Change*. University of Salford and the Centre for Construction Innovation, UK. http://www.scpm.salford.ac.uk/trust

Whitney, J. O. 1996. *The Economics of Trust—Liberating Profits and Restoring Corporate Vitality*. McGraw Hill, New York.

William, S., G. Wood, P. McDermott, and R. Cooper. 2002. Trust in construction: Conceptions of trust in project relationships. In *W92 2002 Conference—Trinidad and Tobago*. http://www.scpm.salford.ac.uk/trust

Wong, E. S., D. Then, and M. Skitmore. 2000. Antecedents of trust in intra-organizational relationships within three Singapore public sector construction project management agencies. *Construction Management and Economics* 18: 797–806.

Wong, P. S. P., and S. O. Cheung. 2004. Trust in construction partnering: Views from parties of the partnering dance. *International Journal of Project Management* 22(6): 437–446.

Wong, P. S. P., S. O. Cheung, K. K. W. Cheung, and C. H. Suen. 2003. Nature and function of trust in construction partnering in Hong Kong. *Proc., Joint International Symposium of CIB W55, W65 and W107.*764–773.

Wood, G., and P. McDermott. 1999. Searching for trust in the UK construction industry: An interim view. In *CIB W92 Procurement Systems Conference*. http://www.scpm.salford.ac.uk/trust

Wright, J. N. 1997. Time and budget: The twin imperatives of a project sponsor. *International Journal of Project Management* 15(3): 181–186.

Co-operation: Good Faith and Beyond

Sai On Cheung

8.1 Introduction

Co-operative contracting in construction, in the form of partnering, has been used to alleviate the problem of confrontation and mistrust in the construction industries of the United States, Australia, and UK. With its positive effects such as reduction in claims and disputes, improved quality, better safety, reduced construction time and cost (CIRC 2001), co-operative contracting is widely regarded as the means to solve some long-standing problems of the construction industry. Co-operative contracting involves seeking win-win outcomes that value long-term relationship. Partnering aims at putting the handshake back into the contracting process (Johnson 1990). It is a collaborative approach to contract management and maximizing the parties' benefits (Charles and Conan 1990). As such, trust and openness are norms, thus enabling an environment for open discussion of problems (Cook and Hancher 1991).

With the increasing use of co-operative contracting in the construction industry, understanding the legal implications of a partnering agreement is of critical importance. One of the risks is the legal enforceability of a co-operative agreement which is either not intended to have legal effect or unenforceable per se. It is treated as a moral contract or agreement in which the parties owe each other reciprocal duties of good faith performance (Barlow 2003, Heal 1999). The underlying spirits thus include co-operation, trust and equality. These are consonant with the concept of good faith (Steven 1993). Colledge (2000) describes undertaking co-operative contracting involves agreeing key mutual objectives to which all parties subscribe. Therefore the "agreement" often includes expressions such as a commitment to co-operation, teamwork or trust. The establishment of a commitment from all the project participants at all levels to make the project a success is therefore pivotal to all co-operative contracting attempts. As such, effective project delivery does not rest on contract as many would advocate. What matters is how the project participants treat each other. Moreover, in an industry that is so entrenched with adversarial attitude, acting cooperatively seems something more of an irony than reality (Colledge 2000). The requirement to co-operate can be expressed as requiring teamwork. More typically, it appears as part of a project common pledge in the form of a partnering agreement. The enforceability of these requirements has been a major concern by those legally minded. Conceptually it is not difficult to appreciate the meaning of co-operation, teamwork or trust. However where legal interpretation is needed, these concepts become too vague to be enforced. The doctrine of good faith is considered to be the most relevant legal ground to be invoked if the agreement to co-operate is not observed. However, unlike the civil law system, the common law system does not recognize a general obligation of good faith. Instead, piecemeal covenants against bad faith are available. The enforcement of the doctrine of good faith in partnering agreements remains uncertain. Having a better understanding of such risk is therefore essential to the implementation of co-operative contracting in the industry. In this chapter a review on the doctrine of good faith, in particular the implementation of co-operative contracting in construction is provided. Some of the key issues are:

(1) To examine the doctrine of good faith in the common law system;

(2) To review the application of the doctrine of good faith in construction contracts, and

(3) To forecast the enforceability and extent of the express/implied obligation of good faith arising from agreements to co-operate.

In has been suggested by Heal (1999) and Colledge (2000) that the spirit of co-operative contracting is akin to the principle of good faith found in civil law jurisdiction. Adams and Brownsword (1995) suggest that the doctrine of good faith is commonly seen as the most direct way of importing into contract law the idea of co-operation. It appears that the answer varies with the jurisdiction involved. For example, Heal (1999) reported that the Canadian courts have moved towards finding a general implied duty of good faith and fair dealing in contracts.

8.2 Meaning of Good Faith

The doctrine of good faith is recognized as one of the general principles of contract law (Brownsword *et al.* 1999) in many legal systems around the world. However, English

Law has been eschewing such an approach (Colledge 2000). Instead fair dealings are secured through the reliance on a number of specific doctrines against bad faith.

The main difficulty associated in accepting a general principle probably lies in the vague meaning of good faith. In fact, interpretation of good faith varies with, the subject matter, the contract type as well as the context. Further difficulties include i) the doctrine of good faith would call for inquiries into the states of mind of the contracting parties; ii) the doctrine of good faith is inconsistent with the fundamental philosophy of freedom of contract; and iii) a general doctrine of good faith does not recognize the contextual variations under which contracts are made (Wightman 1999).

The fact that good faith does not have a universally accepted definition offers little help. Two board definitions of good faith come from the Uniform Commercial Code (UCC) of the United States. These are "honesty in fact in the conduct or transaction concerned" and "honesty in fact and the observance of reasonable commercial standards of fair dealing in the trade". Summers (1968) employed an "excluder" theory which identifies good faith by way of contrast with the specific and variant forms of bad faith which judges decide to prohibit. Table 8.1 summarizes a list of bad faith conduct and their corresponding good faith implications.

Wightman (1999) described three forms of good faith: core, contextual and normative. The core version of good faith refers to the minimum standard of honesty in the formation of contact. An example is the "honesty in fact" as expounded by the definition of good faith in the Uniform Commercial code. The rules on misrepresentation are the analogies in English Law. The notion of honesty in fact is admitted without much controversy.

Contextual good faith derives its meanings from the reasonable expectation of the contracting parties. These expectations are those widely observed in their contracting community, as such are reflective on "standards" being practiced. And it is for this reason that these "standards" are not universal and vary with localities. In English Law, the function of contextual good faith is discharged by the application of implied terms. Wightman (1999) outlined three conditions for the development of "standards" within certain contracting community. Firstly, the type of practice or contacting must be regularly conducted. Secondly, there should not be substantial power difference between the contracting parties. Thirdly, the contract must be relational in the sense that the contracting parties are interdependent and the contracting community has the experience to deal with problems arising from the contract.

Table 8.1   A List of Bad Faith Conduct and Its Corresponding Good Faith Implications

Form of Bad Faith Conduct	Good Faith Implications
Seller concealing a defect in what he is selling.	Fully disclosing material facts.
Builder willfully failing to perform in full, though otherwise substantially performing.	Substantially performing without knowingly deviating from specifications.
Contractor openly abusing bargaining power to coerce an increase in the contract price.	Refraining from abuse of bargaining power.
Hiring a broker and then deliberately preventing him from consummating the deal.	Acting cooperatively.
Conscious lack of diligence in mitigating the other party's damages.	Acting diligently.
Arbitrarily and capriciously exercising a power to terminate a contract.	Acting with some reason.
Adopting an overreaching interpretation of contract language.	Interpreting contract language fairly.
Harassing the other party for repeated assurances of performance.	Accepting adequate assurances.

Source: Summers 1968

Normative good faith essentially is a form of contractual justice which is imposed on the parties. For example, the Restatement (2ed) of contracts states that: "Every contract imposes upon each party a duty of good faith and fair dealing in its performance and its enforcement." As such, the distinction between the contextual and normative approach is that: the contextual approach to good faith involves an interpretation by the court of the meaning of the norms in the parties' contractual community, while the normative approach involves a court applying a standard of good faith drawn from elsewhere (Wightman 1999).

In his paper that won the 1996 Society of Contracts Law Hudson Prize, Bick (1996) outlined how Good Faith is put in practice in the formation of engineering contacts. Following the concepts depicted in contextual good faith, an absence of good faith will have occurred when the standard of commercial behavior displayed by one party towards others falls below that which is accepted as honest within the market place (Bick 1996). In this aspect, a transparent bidding process is an illustrative example of good faith practice whereas "referential" bids, collusive bidding and cover pricing are examples of bad faith practice.

The English Law has yet established a duty of disclosure as far as site and other similar information are concerned. In other jurisdictions, such as Australia, New Zealand and in particular Canada, that give due consideration to the concept of good faith, there is an affirmative duty to disclose.

8.3 Good Faith in Common Law System

Brownsword (1999) summarized four arguments for the acceptance of a doctrine of good faith as follows: "Firstly, it was highlighted that various doctrines have already been established to regulate bad faith dealings, thus adopting a general principle of good faith is just a more rational way to address bad faith dealings more openly and directly. Secondly, by accepting good faith as an umbrella principle that covers, unifies and fills the gaps between a range of specific doctrines designed to secure fair dealing, judges can rely on this umbrella principle to deal with difficult cases, in particular, for one-off decision. Thirdly, in direct response to one of the underpinnings of the negative view, a general principle of good faith will better equip the courts to deal with the varying expectations arising from the many different contracting contexts. Fourthly, the existence of a good faith doctrine would derive the beneficial effect of inducing an environment that gives contracting parties the flexibility and security to do business in a less adversarial orientation." Thus a doctrine of good faith can contribute to the nurturing of a trusting and co-operative contacting environment.

Furthermore, Brownsword (1999) deliberated the negative, neutral and positive reception of good faith in English Contract Law. The negative view suggested that the adoption of a good faith doctrine would be a bad thing. The positive view considers it would be a good thing in adopting a good faith doctrine while the middle ground is the neutral view. The English Contract Law has tended to take the negative view by taking the stance that acting in good faith implies that the contracting parties would exercise restraint in pursuing self-interest and take account of the legitimate interests or expectations of the other party. This attitude is considered as incompatible with the hard headed commercial reality. The other underpinning of the negative view is the moral standard required to exercise good faith is too uncertain and vague. Brownsword (1999) posed the following questions and opined that unless more definitive answer can be provided for these questions, the concerns of good faith skeptics cannot be evaded.

"(1) Whether good faith requires only a clear conscience (subjective good faith) or it imports a standard of fair dealing that is independent of personal conscience (objective good faith)?

(2) Whether good faith applies to all phases of contracting, including pre-contractual conduct?

(3) Whether good faith regulates only conduct (namely, how the parties conduct themselves during the formation of the contract and, subsequently, how they purport to reply on the contractual terms for performance, termination, and enforcement) or also the content (substance) of contracts—in other words, whether good faith regulates matters of procedure and process or also matters of contractual substance?

(4) Whether a requirement of good faith adds anything to the regulation of bad faith—whether good faith simply comprises so many instances of bad faith (as Robert Summers famously argued) or whether following Lon Fuller's terminology (albeit not in this context) bad faith sets a minimal morality of duty while good faith reflects a morality of aspiration?

(5) Whether good faith imposes both negative and positive requirements (covering, say, non-exploitation, non-opportunism, non-shirking as well as positive co-operation, support, and assistance)?" (Brownsword 1999)

The neutral view takes the stand that there is nothing intrinsically objectionable about a good faith doctrine, nevertheless, English law already has its doctrinal tools such as economic duress, promissory estoppels, misrepresentation, mistake, frustration and the like to guard against unfair dealings. Principally this attitude can be identified as the "equivalence thesis" or the "indifferent thesis" (Brownsword 1999). Moreover, it is argued that the neutral view is inclined towards the negative view for the simple reason that the choice between a new concept as in the case of introducing a doctrine of good faith and applying established doctrines, albeit piecemeal, is rather obvious.

In sum, there is no general obligation to observe good faith in Common Law (O'Connor 1990). As discussed in Chapter three, the rationale behind this approach is the principle of freedom of contract accorded in Common Law where intervention in contract is only permitted as exception (Groves 1999). As such, mastering the principle of freedom of contract will help to understand the roles of good faith or guards against bad faith in Common Law. The overriding principle of freedom of contract can in fact be divided into two different but related forms. According to Cohen (1995), the first form is a positive one, which means that the parties are free to create a binding contract and make the terms of their agreement. The second form is a negative one meaning that the parties are free from obligations so long as a binding contract has not been concluded. It can be seen that the first positive freedom operates at the time of creation and performance of contract whereas the negative is relevant to the pre-contract period.

8.3.1 Good Faith in Formation of Contract

One of the classical approaches of the common law courts is the freedom of formation of contract at the pre-contractual period (Groves 1999), that is, there is no obligation between the parties prior to the formation of a contract (Cohen 1995). However, a strict application of the doctrine of freedom of contract would lead to injustice in certain circumstances. That is why the courts occasionally would intervene to govern pre-contractual conduct of parties (Groves 1999). The courts have used various techniques or principles such as duty not to act in bad faith, implied contract, unjust enrichment, equitable estoppel and misrepresentation to do justice (Groves 1999). Although good faith is not explicitly recognized in the pre-contractual stage, good faith can be considered as the underlying motivation to these seemingly disparate exceptions to freedom of contract (Groves 1999).

8.3.2 Good Faith in Performance of Contract

When the parties have entered into a formal contract, it is the positive form of freedom that comes into play (Whittaker and Zimmermann 2000). This freedom allows contracting parties entering into a binding contract and making the terms of their agreement and be bound by those terms (Cohen 1995). Whittaker and Zimmermann (2000) pointed out that parties are entitled to exercise their rights arising under the contract or under the law of breach of contract for whatever reason they choose. For example, in the case of *James Spencer & Co. Ltd. v. Tame Valley Padding Co. Ltd.* (1998), Potter L. J. held that "the plaintiffs are free to act as they wish provided that they do not act in breach of a term of the contract". An extreme example of the irrelevance of motive to the exercise of a right arising under a contract may be found in the decision of the Court of Appeal in *Chapman v. Honig* (1963). In this case, Pearson L. J. held that "A person who has a right under a contract or other instrument is entitled to exercise it and can effectively exercise it for a good reason or no reason at all". These examples suggest that English Law does not have a general requirement of good faith in the performance of contract and that reflects the tradition of freedom of contract.

However, contracting parties may not always enjoy complete positive freedom of contract. In some circumstances the courts may wish to impose restrictions. For example, absolute exercise of contractual right is limited by the doctrine of frustration (Whittaker and Zimmermann 2000). Another curtailment of freedom of contract is the courts' control on remedies (Friedmann 1995). In *Dunlop Pneumatic Tyre Company*

Limited v. New Garage and Motor Company Limited (1915), the court held that it would strike out agreed damages clauses where these constitute a penalty rather than a genuine pre-estimate of the loss likely to be suffered following a particular breach. Thus, the courts will withhold a particular contractual remedy should a party had acted in bad faith (Groves 1999). In addition, there are a number of specific rules which limit the possibilities of unfair or unreasonable termination (Friedmann 1995). The doctrine of substantive performance is an example. Under this doctrine the party in breach, who has substantially performed the work, is entitled to recover the contractual price. Such doctrine alleviates the harshness of agreed contract terms and prevents one party from taking advantage by strictly sticking to the terms of contract.

8.4 Good Faith in Action in Construction

According to Colledge (1999), English law has embraced implicitly the following concepts of good faith in construction contracts:

Implied Terms of Co-operation

In construction contracts, the relationship between the client and the contractor is often described as co-operative (Wallace 1986). These constitute two general implied terms in construction contracts: (i) duties not to obstruct; and (ii) duties to co-operate actively (Burrows 1968).

Duties Not to Obstruct

These duties imply that neither party shall do anything to hinder the other from performing the contract (Colledge 1999). For instance, the employer is not entitled to claim liquidated damage for the delay caused by his own; and the employer cannot interfere with architect's duty in issuing certificates. Burrows (1968) illustrated through a group of cases which held that an employer was not entitled to enforce the liquidated damage clause if the delay of the contract was caused by the employer himself. Many standard forms of construction contract require certification by the architect or engineer on various issues such as payment, the date of completion or the date of making good defects etc. Such certificates affect the rights and obligations available for the parties under the construction contract. For example, the employer can only deduct liquidated damages if the architect issues a certificate of non-completion; and the contractor cannot demand payment without the architect or engineer's payment certificates. In exercising this certification function, the architect or engineer must act honestly and in good faith, even though the architect or engineer are the agents of the employer (Groves 1999).

Duty to Co-operate Actively

Another general implied duty is the duty to co-operate actively (Burrows 1968). It is an obligation to do thing which is necessary to enable the other party to perform his or her obligations (Colledge 1999). An example of such implied duty in construction is the provision of necessary information and drawings to the contractor within reasonable time by the employer. The employer has the obligation to provide information to the contractor within a reasonable time to be determined from the express terms of the contract and all circumstances, including the views of architect/engineer and employer. In other words, there is a positive duty to provide information at a time, so as not to hinder or prevent the contractor from completing the works in accordance with the contract. However, the courts seek to limit the extent of a positive co-operation duty and will not impose obligations which go beyond those contemplated by the parties (Colledge 1999).

8.5 Enforcement of Good Faith Obligation in Co-operative Contracts

In both Australia and the UK, it appears that the courts are moving towards explicit recognition of good faith obligations in the performance of contract. The likelihood of the court to enforce the express and implied terms of good faith in partnering contracts and the development of good faith obligation are examined in this section.

8.5.1 Express Term of Good Faith

Express terms of good faith are incorporated into co-operative contracts by either binding charters or express good faith provisions. It clearly reflects the parties' intention to have the duty to be legally binding unless expressed otherwise. For the purposes of maintaining business efficacy, and respecting freedom of contract, the express obligations of good faith are likely to be upheld by the courts unless the express duties of good faith lacks certainty. A notable example is when the express term of good faith contradicts the surrounding circumstances or other express terms (Colledge 2000). Moreover, two issues arising from express terms of good faith need to be elaborated. Firstly, whether the term will affect the interpretation of other contract provisions and, secondly, what extent of obligations will be created by the express term.

Regarding the first issue, Butcher (1997), Helps (1997) and Critchlow (1998) considered such express terms of good faith are having impact on the interpretation of other contract provisions and could result in enhanced obligations of co-operation being placed on the parties. For example, in the case of GC/Works/1 (1998) Contract in UK, Clause 56 of the contract termination provision would be subject to the good faith requirement set out in the Clause 1A(1). Although Clause 56(8) permits termination by the employer at his will by notice to the contractor and no ground for determination is required, due to the effect of introducing the express term of good faith, the use of this termination power might be required to be exercised in good faith. The relevant clauses are as follow:

> Clause 56(8): "Without prejudice to any other power of determination, the Employer may at will determine the Contract by notice to the Contractor . . . "

> Clause 1A(1): "The Employer and the Contractor shall deal fairly, in good faith and in mutual co-operation, with one another, and the Contractor shall deal fairly, in good faith and in mutual co-operation, with all his subcontractors and suppliers."

Besides whether express term of good faith would affect the interpretation of other terms, the more uncertain issue is: What might the express obligation of good faith be extended? Brownsword (1994) suggested two examples. One is described as "good faith as an exception" which means that the contracting parties may legitimately pursue their own interests, prioritizing their own interests against those of the other side, subject only to minimal constraints such as fraud and coercion. The second is referred as "good faith as the rule" which means that each party must respect the legitimate interests of the other contracting party. In other words, the contracting parties should attempt to promote their joint interests and they may not legitimately prioritize their own interests against the protected interests of the other side (Brownsword 1994). The former is a concept of good faith which prevents opportunistic behavior. The latter is a concept which requires the extension of good faith principle beyond general duties of honesty or fair dealing.

In contracts, where there is express provision of good faith, the existence of an agreement to co-operate would provide further evidence to show the co-operative relationship between the parties. In such circumstances, obligations of positive co-operation might be expected and enforced (Helps 1997). Therefore, Brownsword (1994) suggested that the courts should look positively at good faith obligations if the partnering charter is binding and the contract includes an express good faith provision.

8.5.2 Implied Term of Good Faith

In the absence of express terms of good faith in the contract, whether good faith obligations will be upheld is less certain (Butcher 1997, Helps 1997, Critchlow 1998). However, in construction contracts, the courts are willing to impose the implied duty of co-operation (Colledge 2000). As such, implication of implicit good faith obligation does exist, but the extent of the co-operation duty is limited. As the contracting parties are seldom required to do a positive act which is not expressly stated in the contract. In such situations, they need only to do such a positive act if there is something approaching a necessity for the act to render the express parts of the contract workable

(Burrows 1968). For example, the employer in construction contract is required to provide in due time necessary drawing and information to enable the contractor to discharge its obligations. However, the employer's duty to co-operate does not extend to requiring the employer to assist the contractor to achieve an earlier completion date or to accelerate to make up time lost due to the default of the Employer in his agents.

Regarding the extent of implied obligation of good faith, there is strong support for the courts to uphold negative implied obligation of good faith by reference to the general standards of honesty, fair dealing and reasonableness. With the existence of an agreement to co-operate, it might incentivise the court to uphold the negative obligation of good faith such as an obligation not to hinder or prevent.

For positive implied obligation of good faith, it seems to be more difficult to be established. The courts have been very cautious about the implications arising from imposing a positive duty. It seeks to place limits on the extent of a positive duty and will not impose any obligation which goes beyond that contemplated as constructed by the contract or the intentions of the parties (Colledge 1999). Colledge (1999) provided a good summary of the cases in this connection. In *London Borough of Merton v. Stanley Hugh Leach* (1985), the court held that contracting parties have a duty to take all steps reasonably necessary to enable the others to discharge their obligations, but it was interpreted by the court in *Allridge (Builders) Ltd. v. Grandactual Ltd.* (1997) that the duty means steps which are within the power of the contracting parties. Thus the employer has the obligation to ensure that his representative performs the duties set out in the contract or to give possession of the site to the contractor in accordance with the contract dates, but broader obligations are difficult to be established. Similarly, in *Bedfordshire County Council v. Fitzpatrick Contractors Ltd* (1998), an implied term that a contractor owed to the employer an obligation of trust and confidence was not upheld by the court because the term was considered too imprecise. The main reason why the courts are reluctant to impose positive obligation of good faith is that no guidance had been provided on the interpretation of such terms and it is difficult to establish the degree of assistance required or the extent to which parties are required to have regard to the legitimate interests of each other (Colledge 2000). In other words, the parties' reasonable expectations of positive obligation of good faith are difficult to be ascertained. Where the parties' reasonable expectations cannot be ascertained, the narrower standard of fair dealing or honesty will be implied instead (Colledge 2000). In sum, it seems that there is little chance for the implied obligation of good faith being extended to require positive duties.

8.6 Summary

A growing use of co-operative contracting such as partnering in the construction industry in Hong Kong is evidenced. In fact, the adoption of partnering has already become a new trend (Black 2005). The enforcement of partnering agreement (agreement to co-operate in the context of co-operative contracting) will become an issue that must be addressed. The essential elements of partnering, such as trust, equity and co-operation, are consonant with the doctrine of good faith. The principle of good faith is thus considered as the legal basis upon which the contracting parties seek to gain protection under a partnering arrangement. The obligation of good faith may be incorporated into contracts as express terms by legally binding charter or by good faith provisions, as implied terms by the fact or law, or as terms in collateral contracts. However it is not easy to reduce the meaning of the doctrine of good faith in simple terms. The doctrine in essence requires the contracting parties to impose constraint on its pursuit of self-interest and respect one another's legitimate interests. In the common law system, there is no general obligation to observe good faith in the formation or performance of a contract as compared with those in the civil law system. Instead, varies rules or techniques serving as substitutes of good faith are adopted by the common law courts for achieving justice and fair results. It is in fact an implicit recognition of the doctrine of good faith. The instances of such implicit obligations of good faith can be found in construction contracts, such as an implied term of not to obstruct. However, the implicit obligation of good faith only requires the parties not to act in bad faith. Therefore, there is a distinction between the implicit obligations of good faith and explicit general principle of good faith which may extend the parties' obligations to act pro-actively for the interest of other parties.

In co-operative contracting based projects, if the contract contains an express good faith provision or includes an agreement to co-operate, it is suggested that explicit good faith obligation is very likely to be upheld by the courts. However, to what extent the good faith obligation will be upheld remains uncertain. It appears that a general support from the court in restraining bad faith opportunistic behavior, a result that can also be achieved by the implicit obligation of good faith. With regard to the positive good faith obligation, whether it should be imposed or not, or, how far such obligation should be extended would depend on the surrounding circumstances of the contract.

References

Adans, J., and R. Brownsword. 1995. *Key Issues in Construct*. London: Butterworths.

Barlow, J. 2003. *Partnering, Lean Production and the High Performance Work Place*. School of Construction Housing and Surveying, University of Westminster London. http://web.bham.ac.uk/d.j.crook/lean/iglc4/barlow/barlow.htm.

Bick, P. 1996. Some aspects of good faith and fairness in the formation of construction and engineering contracts. Society of Construction Law.

Black, R. 2005. MTR Partnering—A work in progress. In *CII-HK Conference 2004 on Construction Partnering, 9 December 2004, Hong Kong*.

Brownsword, R. 1994. Two concepts of good faith. *Journal of Contract Law* 7(3): 197.

Brownsword, R. 1999. *Positive, Negative, Neutral: The Reception of Good Faith in English Contract Law in Good Faith in Contract Concept and Context*. Ed. R. Brownsword, N. J. Hird, and G. Ashgate Howells.

Brownsword, R., N. J. Hird, and G. Howells. 1999. *Good Faith in Contract: Concept and Context*. Ashgate.

Burrows, J. F. 1968. Contractual co-operation and the implied term. *Modern Law Review* 31: 390.

Butcher, T. 1997. Partnering: Contractual considerations. *Construction Law* 9(3): 79.

Charles, C. and E. Conan. 1990. A strategy for partnering in the public sector. Commander Portland District, US Army Corps of Engineers.

Cohen, N. 1995. Pre-contractual duties: Two freedoms and the contract to negotiate. In *Good Faith and Fault in Contract Law,* ed. J. Beatson and D. Friedmann. Oxford: Clarendon Press.

Colledge, B. 1999. Good faith in construction contracts—The hidden agenda." *Construction Law Journal* 15(3): 288–299.

Colledge, B. 2000. Obligations of good faith in partnering of UK construction contracts. *The International Construction Law Review* 17(1): 175–201.

Construction Industry Review Committee of Hong Kong (CIRC). 2001. *Construct for Excellence.* Report of the Construction Industry Review Committee. 52–85.

Cook, E. L., and D. E. Hancher. 1991. Partnering: Contracting for the future. *Journal of Management in Engineering* 103(6): 431–446.

Critchlow, J. 1998. We don't need a contract, we're partnering. *Construction Law* 9(6): 183.

Friedmann, D. 1995. Good faith and remedies for breach of contract. In *Good Faith and Fault in Contract Law,* ed. J. Beatson and D. Friedmann. Oxford: Clarendon Press.

Groves, K. 1999. The doctrine of good faith in four legal system. *Construction Law Journal* 15(4): 265.

Heal, A. J. 1999. Construction partnering: Good faith in theory and practice. *Construction Law Journal* 15(3): 167–198.

Helps, D. 1997. Why partnering is not a duty. *Building* 28: 37.

Johnson, D. P. 1990. Senior Counsel for Contracting and Environmental Compliance US Army Corps of Engineers Portland District. Public Section Partnering.

Latham, M. 2001. Engineer briefing note. *The Institution of Civil Engineers.* http://www.ice. org.uk/ rtfpdf/BS-Partnering.rtf.

O'Connor, J. F. 1990. *Good Faith in English Law.* Aldershot: Dartmouth Publishing Company Ltd.

Steven, D. 1993. Partnering and value management. *The Building Economist* (September): 5–7.

Summers, R. S. 1968. Good faith in general contract law and the sales provisions of the uniform commercial code. *Virginia Law Review* 54: 195.

Wallace, I. N. D. 1986. *Construction Contracts: Principle and Policies in Tort and Contract.* London: Sweet & Maxwell.

Whittaker, S., and R. Zimmermann. 2000. Good faith in European contract law: Surveying the legal landscape. In *Good Faith in European Contract Law,* ed. R. Zimmermann and S. Whittaker. Cambridge: Cambridge University Press.

Wightman, J. 1999. *Good Faith and Pluralism in the Law of Contact in Good Faith in Contract—Concept and Context.* Ed. R. Brownsword, N. J. Hird, and G. Ashgate Howells.

Index